Human Anatomy and Physiology

The Forebrain

Development, Physiology and Functions

HUMAN ANATOMY AND PHYSIOLOGY

Additional books and e-books in this series can be found on Nova's website under the Series tab.

HUMAN ANATOMY AND PHYSIOLOGY

THE FOREBRAIN

DEVELOPMENT, PHYSIOLOGY AND FUNCTIONS

MORTEN F. THORSEN
EDITOR

Copyright © 2020 by Nova Science Publishers, Inc.

All rights reserved. No part of this book may be reproduced, stored in a retrieval system or transmitted in any form or by any means: electronic, electrostatic, magnetic, tape, mechanical photocopying, recording or otherwise without the written permission of the Publisher.

We have partnered with Copyright Clearance Center to make it easy for you to obtain permissions to reuse content from this publication. Simply navigate to this publication's page on Nova's website and locate the "Get Permission" button below the title description. This button is linked directly to the title's permission page on copyright.com. Alternatively, you can visit copyright.com and search by title, ISBN, or ISSN.

For further questions about using the service on copyright.com, please contact:
Copyright Clearance Center
Phone: +1-(978) 750-8400   Fax: +1-(978) 750-4470   E-mail: info@copyright.com.

NOTICE TO THE READER

The Publisher has taken reasonable care in the preparation of this book, but makes no expressed or implied warranty of any kind and assumes no responsibility for any errors or omissions. No liability is assumed for incidental or consequential damages in connection with or arising out of information contained in this book. The Publisher shall not be liable for any special, consequential, or exemplary damages resulting, in whole or in part, from the readers' use of, or reliance upon, this material. Any parts of this book based on government reports are so indicated and copyright is claimed for those parts to the extent applicable to compilations of such works.

Independent verification should be sought for any data, advice or recommendations contained in this book. In addition, no responsibility is assumed by the Publisher for any injury and/or damage to persons or property arising from any methods, products, instructions, ideas or otherwise contained in this publication.

This publication is designed to provide accurate and authoritative information with regard to the subject matter covered herein. It is sold with the clear understanding that the Publisher is not engaged in rendering legal or any other professional services. If legal or any other expert assistance is required, the services of a competent person should be sought. FROM A DECLARATION OF PARTICIPANTS JOINTLY ADOPTED BY A COMMITTEE OF THE AMERICAN BAR ASSOCIATION AND A COMMITTEE OF PUBLISHERS.

Additional color graphics may be available in the e-book version of this book.

Library of Congress Cataloging-in-Publication Data

Names: Thorsen, Morten F., editor.
Title: The forebrain: : development, physiology and functions / Morten F.
   Thorsen, editor.
Description: New York : Nova Science Publishers, [2020] | Series: Human
   anatomy and physiology | Includes bibliographical references and index.
Identifiers: LCCN 2020032554 (print) | LCCN 2020032555 (ebook) | ISBN
   9781536184075 (paperback) | ISBN 9781536184266 (adobe pdf)
Subjects: LCSH: Prosencephalon. | Cholinergic mechanisms.
Classification: LCC QP382.F7 F67 2020 (print) | LCC QP382.F7 (ebook) |
   DDC 612.8/25--dc23
LC record available at https://lccn.loc.gov/2020032554
LC ebook record available at https://lccn.loc.gov/2020032555

Published by Nova Science Publishers, Inc. † New York

Contents

Preface vii

Chapter 1 Building the Caudal Forebrain Signaling Center 1
Béatrice C. Durand

Chapter 2 Contribution of the Basal Forebrain Nuclei
to the Control of Cortical Activity in Rodents 61
Irene Chaves-Coira, Jesús Martín-Cortecero,
Margarita L Rodrigo-Angulo and Angel Nuñez

Chapter 3 The Role of Reelin in Cortex-Dependent
Motor Function 99
Mariko Nishibe and Yu Katsuyama

Index 125

PREFACE

In this compilation, the authors provide an overview of neural plate pre-patterning and the concept of the organizing center, describing the contributions of informative cues and signaling pathways involved in zli positioning.

Continuing, The Forebrain: Development, Physiology and Functions aims to show the existence of specific neuronal populations in basal forebrain linking with specific sensory, motor and prefrontal cortices. In addition, the electrophysiological properties of cholinergic pathways that control cortical activity are examined.

In closing, the authors discuss the putative involvement of Reelin signal in motor-related impairments observed in neurological diseases, including lissencephaly, psychiatric disorders and brain injuries.

Chapter 1 - Development of the caudal part of the forebrain from which emerges the thalamus and the habenula relies on the morphogen Sonic Hedgehog (Shh) whose activity participates in the survival, growth and patterning of specific neuronal progenitor subpopulations. Besides the neural tube basal plate, Shh is secreted by a small group of specialized cells uniquely located in the dorsal part of the neural tube: the *zona limitans intrathalamica (zli)*. The *zli* is the last brain-signaling center to form and the first forebrain compartment to be established. In this chapter the author describes the principles underlying emergence of this unique brain organizer. The author presents an overview or neural plate pre-patterning and the concept of organizing center. The author describes the contributions of informative cues and signaling pathways involved in *zli* positioning. The author highlights the network of Transcription Factors (TFs) and *cis*-regulatory sequences underlying formation of a shh-expressing delimitated cell compartment in the anterior brain. The author discusses evidence that this network predates the origin of chordates and the difference of *zli* developmental modes observed between species. The author shortly discusses implications for the fact that Shh relies on primary cilia for signal transduction and give some elements on the etiology of brain malformations associated to defects in diencephalic early development and Shh signal transduction.

Chapter 2 - Numerous evidences support that specific projections between different basal forebrain (BF) nuclei and their cortical targets are necessary to modulate cognitive functions in the cortex. These nuclei provide most of the cholinergic innervation to sensory, motor and prefrontal cortices, and hippocampus. Data suggest that these BF nuclei are integrated in distinct BF-cortical networks that may play different roles in sensory processing, motor control, or cortical arousal. The aim of this review is to show the existence of specific neuronal populations in the BF linking with specific sensory, motor and prefrontal cortices. In addition, electrophysiological properties of cholinergic pathways to control cortical activity are shown. Finally, the results obtained mainly in rodents could explain some aspects that are observed in neurodegenerative diseases.

Chapter 3 - *Reelin* has been identified as a gene responsible for the abnormal phenotype observed in *reeler*, the classic mouse mutant strain that exhibits some functional impairments reminiscent of these found in neurological disorders. Reelin is an extracellular glycoprotein which regulates and comprises intracellular biochemical signaling through binding to its cell membrane receptors. Studies undertaken in the last two decades, using human brain specimens and human genome data, have suggested the involvement of *REELIN* in various brain pathologies. Early reports on abnormalities in *reeler* described ataxia attributed to cerebellar dysfunction and disruption of laminar structure in the cerebral cortex. More recent studies on *reeler*, on the other hand, have described detailed abnormalities of the cortex at the level of synaptic functions, neuronal plasticity and dendritic spine formation. In this review, the authors discuss the putative involvement of Reelin signal in motor-related impairments observed in neurological diseases, including lissencephaly, psychiatric disorders and brain injuries.

In: The Forebrain
Editor: Morten F. Thorsen

ISBN: 978-1-53618-407-5
© 2020 Nova Science Publishers, Inc.

Chapter 1

BUILDING THE CAUDAL FOREBRAIN SIGNALING CENTER

Béatrice C. Durand[*], PhD
Developmental Biology Laboratory, Sorbonne Université, Paris, France

ABSTRACT

Development of the caudal part of the forebrain from which emerges the thalamus and the habenula relies on the morphogen Sonic Hedgehog (Shh) whose activity participates in the survival, growth and patterning of specific neuronal progenitor subpopulations. Besides the neural tube basal plate, Shh is secreted by a small group of specialized cells uniquely located in the dorsal part of the neural tube: the *zona limitans intrathalamica (zli)*. The *zli* is the last brain-signaling center to form and the first forebrain compartment to be established. In this chapter I describe the principles underlying emergence of this unique brain organizer. I present an overview or neural plate pre-patterning and the concept of organizing center. I describe the contributions of informative cues and signaling pathways involved in *zli* positioning. I highlight the network of Transcription Factors (TFs) and *cis*-regulatory sequences underlying formation of a shh-expressing delimitated cell compartment in the anterior brain. I discuss evidence that this network predates the origin of chordates and the difference of *zli* developmental modes observed between species. I shortly discuss implications for the fact that Shh relies on primary cilia for signal transduction and give some elements on the etiology of brain malformations associated to defects in diencephalic early development and Shh signal transduction.

Keywords: developmental biology, embryogenesis, neural, segregation, compartment, sonic hedgehog, forebrain, thalamus, holoprosencephaly, patterning

INTRODUCTION

Brain development is a highly sophisticated and immensely complex process that has intrigued biologist for centuries. Observations and functional experiments on different model systems provided crucial insights into the early molecular, cellular, and morphological events underlying this fascinating process. Such studies revealed the importance of so called "signaling centers": small groups of specialized cells acting as local sources of secreted morphogens. These secreted morphogens act on survival, proliferation, and fate commitment of neuroepithelial cells. Thereby they organize, and pattern, the compartmentalization of the neuroepithelium into functional histological units from which emerge the different brain parts sustaining its complex function (Jessell and Sanes 2000, Hebert and Fishell 2008, Scholpp and Lumsden 2010, Cavodeassi and Houart 2012, Kiecker and Lumsden 2012, Echevarri et al. 2003, Anderson and Stern 2016).

The most anterior part of the neuroepithelium, the prosencephalic neural plate will generate the forebrain (also called prosencephalon). The forebrain is divided into two parts: most anteriorly the telencephalon that develops into the cerebral cortex, and the posterior or caudal part, the diencephalon that is located between the cerebral cortex and the midbrain. From the diencephalon emerges the thalamic complex, a bilateral structure that contains the pre-thalamus (ventral thalamus) rostrally, and the thalamus (dorsal thalamus) caudally. The epithalamus rests dorsally on the thalamus, and generates the habenula. In vertebrates the dorsal thalamus, and the habenula, are at the core of neuronal connections between the forebrain limbic system, and the midbrain and/or hindbrain. They thereby are functionally crucially important in the modulation of emotion, motivation and rewards (Scheibel 1997, Huang et al. 2018, Shelton, Becerra, and Borsook 2012).

Emergence of the thalamus complex is orchestrated by a brain signaling center called the *zona limitans intrathalamica,* or *interthalamica* (*zli*), also referred to as the mid-diencephalic organizer (MDO). The *zli* is defined by the alar plate expression of Sonic HedgeHog (Shh), a secreted morphogen that mediates regionalization of the prethalamus anteriorly and the thalamus posteriorly (Scholpp et al. 2006, Vieira et al. 2005, Juraver-Geslin, Gomez-Skarmeta, and Durand 2014). Shh signaling secreted from the *zli*, participates in the survival, growth, and patterning of neuronal progenitor subpopulations within the thalamic complex and the habenula (Scholpp et al. 2006, Vieira et al. 2005, Kiecker and Lumsden 2004, Hashimoto-Torii et al. 2003, Juraver-Geslin, Gomez-Skarmeta, and Durand 2014, Epstein 2012, Jeong et al. 2011, Szabo et al. 2009, Chatterjee et al. 2014).

In this chapter we will present an overview or the informative cues patterning the neuroepithelium and the concept of organizing center. As the activity of Shh in diencephalic patterning has previously been detailed (Chatterjee and Li 2012, Epstein 2012, Scholpp and Lumsden 2010, Martinez-Ferre and Martinez 2012), we here focus on the principles underlying building of the *zli*. We highlight the contributions of early cues and signaling pathways in *zli* formation. We describe the network of transcription factors and *cis*-regulatory sequences that allow competence for *zli* establishment. We describe the role of Transcription Factors (TFs) on emergence of the *zli* as a cell compartment boundary within the forebrain territory. We shortly discuss implications for the fact that Shh relies on primary cilia for signal transduction and give some elements on the etiology of brain malformations associated to defects in diencephalic early development and Shh signal transduction.

INFORMATIVE CUES IN EARLY PATTERNING OF THE NEUROEPITHELIUM

The Neuroepithelium: A Checkerboard of Histogenic Fields

In vertebrates the largest part of the nervous system comes from a homogenous layer of neuroepithelial cells: the neural plate. Whereas the ontology of our brain still displays a lot of mystery, it is now known that the neural plate subdivides into distinct neural territories through simultaneous and sequential steps. Based on morphological considerations in different species, embryological manipulations, and the expression pattern of regulatory genes, a conceptual model holds that the neural plate is divided into transverse and longitudinal segments that define a developmental grid generating distinct histogenic fields. Some fields then become compartmentalized, specified by a unique pattern of gene expression and maintained by polyclonal cell lineage restriction. Several pioneering teams performed elegant fate-mapping experiments in zebrafish (Woo, Shih, and Fraser 1995), Xenopus (Eagleson and Harris 1990) and chick (Couly and Le Douarin 1987, Figdor and Stern 1993, Lumsden and Keynes 1989). These fate map analyses demonstrate that the position of the histogenic fields in the neural plate is a flat representation of their topological relationships in the mature neural tube. Thereby the primordia of the forebrain, midbrain, hindbrain and spinal cord are all already established along the antero-posterior (AP) axis of the neural plate (reviewed in (Hoch, Rubenstein, and Pleasure 2009, Heasman 2006, Wilson and Houart 2004)). In consequence a construction blueprint of the neural organization, and specifically that of the forebrain, is established during neurulation. At the neural plate stage when neural induction is occurring, the histogenic fields can be detected as spatially restricted domains of gene expression (Puelles and Rubenstein 2003, Rubenstein et al. 1994). Analysis of these patterning cues has unveiled some of the genes regulatory networks (GRNs) involved in the specification of neural territories (reviewed in (Beccari, Marco-Ferreres, and Bovolenta 2013)). Correct development of a histogenic field depends on two critical biological processes that act in parallel and synchronously: the specification of neuroepithelial cells into distinct cell types, and the growth of the neuroepithelium (reviewed in (Dreesen and Brivanlou 2007)).

Informative Cues for Neuroepithelial Patterning and Growth

The initial regionalization of the neuroepithelium relies on synergistic activities of at least five major signaling factors secreted first by the blastopore lip organizer, and second by the axial mesoderm, that both signal in a vertical and planar manner (reviewed in (Niehrs 2004, Stern 2002)). The combinatorial activity of secreted signals both induces the expression of identity genes, and modulates the growth of the different sub-regions of the central nervous system (CNS). Experiments mainly performed in amphibian, avian and zebrafish embryos established roles for bone morphogenetic proteins (BMPs), fibroblast growth factors (FGFs), Hedgehog (HH), Wnt, and Nodal proteins in neural induction, neural proliferation, and establishment of the future Dorso-Ventral (DV) and Antero-Posterior (AP) axes of the nervous system. Early roles of these signaling pathways are highly conserved during evolution (Beddington and Robertson 1999, Jessell and Sanes 2000). Inhibition of the BMP pathway is involved in the decision of ectodermal cells to become epidermal or neural cells. Sonic Hedgehog (Shh), help regionalize the DV axis of the developing nervous system in a concentration-dependent way. Wnt signaling initially participates in the establishment of a crude AP axis and later, together with FGFs, participates in a more refined patterning of the anterior neural tube (reviewed (Stern 2005, Wilson and Rubenstein 2000)). Besides instructing cell fates, signaling factors control the growth of the neural tissue, a process that depends on a delicate balance between the proliferation of neural progenitor cells and Early Programmed Cell Death (EPCD), primarily through apoptosis (reviewed in (Raff 1998)). BMPs, FGFs, Wnt and Shh are all thought to be involved in controlling proliferation and apoptosis in the developing nervous system ((Chenn and Walsh 2002, Megason and McMahon 2002) reviewed in (Mehlen, Mille, and Thibert 2005)).

The Signaling Centers

Local sources of secreted patterning factors emerge during neural development. These small groups of specialized cells fulfill a localized patterning function thereby acting as "organizing centers." In 1924, the first, and for some the only, "organizing" center was discovered by Hans Spemann and Hilde Mangold. The two scientists demonstrated that the dorsal lip of the blastopore, if grafted into the ventral part of a host embryo, induces a secondary axis containing a complete nervous system (reviewed in (De Robertis et al. 2000, Niehrs 2004, Stern 2001)). Since this discovery in amphibian, this organizing center has been referred to as "Spemann organizer," and further identified in other model organisms; Hensen's node in the chick, the node in the mouse and the shield in zebrafish. Moreover emergence of a blastopore-associated axial organizer has been demonstrated to predates the cnidarian-bilaterian split over 600 million years ago (Kraus et al. 2016). This primary organizer is the first tissue to differentiate during embryonic development. It emerges at gastrulation and gives rise to the prechordal plate and the notochord, two tissues that send planar and vertical signals to the overlying prospective neuroepithelium (reviewed in (Stern 2002, Wilson and Houart 2004, Niehrs 2004)). Through the secretion of BMP signaling pathway inhibitors, the primary organizer induces the neuroectoderm from the dorsal ectoderm, which gives rise to the neural plate (Harland 2000). Of note interactions between BMP and other signaling pathways such as Wnt and FGF are important in this process (Stern 2006). Secondary signaling centers emerge later in development. These signaling centers are of fundamental importance for the regionalization of the brain, since they allow structures of the CNS to refine their development (Cavodeassi and Houart 2012, Echevarri et al. 2003, Kiecker and Lumsden 2012). Noteworthy they are always located at molecular frontiers separating territories. The anterior neural ridge (ANR) borders the neural plate anteriorly at the interface between the non-neural ectoderm and the neuroectoderm. The ANR expresses Fgf8 and specifies the forebrain (Houart, Westerfield, and Wilson 1998). The midbrain-hindbrain boundary (or isthmic organizer) separates the anterior and posterior neural tube and secretes Wnt1 and Fgf8 (Crossley and Martin 1995). The floor plate through the secretion of Shh, and the roof plate through that of Wnts and BMPs ligands also act as signaling centers (reviewed in (Echevarri et al. 2003, Alexandre and Wassef 2003, Cavodeassi and Houart 2012)). Finally, the *zona limitans intrathalamica* (*zli*) that is the focus of the present chapter, presents itself as a dorso-ventrally (DV) extending gene expression domain of the morphogen Shh. It is the latest signaling center to emerge and the only part of the dorsal neural tube expressing Shh (Scholpp and Lumsden 2010, Scholpp et al. 2006, Vieira et al. 2005, Kiecker and Lumsden 2004, Hashimoto-Torii et al. 2003, Juraver-Geslin, Gomez-Skarmeta, and Durand 2014, Epstein 2012, Jeong et al. 2011, Szabo et al. 2009, Chatterjee et al. 2014).

CHARACTERISTICS OF A CAUDAL FOREBRAIN SHH-EXPRESSING COMPARTMENT

Zli Position within the Forebrain Map

Based on morphology (Bergquist and Kallen 1954, Coggeshall 1964, Keyser 1972, Vaage 1969) and gene expression (Bulfone et al. 1993, Rubenstein et al. 1994), the diencephalic primordium is divided into three transverse segments called prosomeres (p) that generate three distinct histogenic fields: p3, or anterior parencephalon, corresponds to the presumptive prethalamus and the eminentia thalami; p2, or posterior parencephalon, gives rise to the epithalamus and the thalamus; and p1, or synencephalon, generates the presumptive pretectum. Each prosomere is divided into a ventral (basal) and a dorsal (alar) part (Figure 1) (reviewed in (Figdor and Stern 1993, Puelles and Rubenstein 2003, Martinez-Ferre and Martinez 2012, Thompson et al. 2014, Ferran, Puelles, and Rubenstein 2015, Puelles 2018, Puelles et al. 2013)).

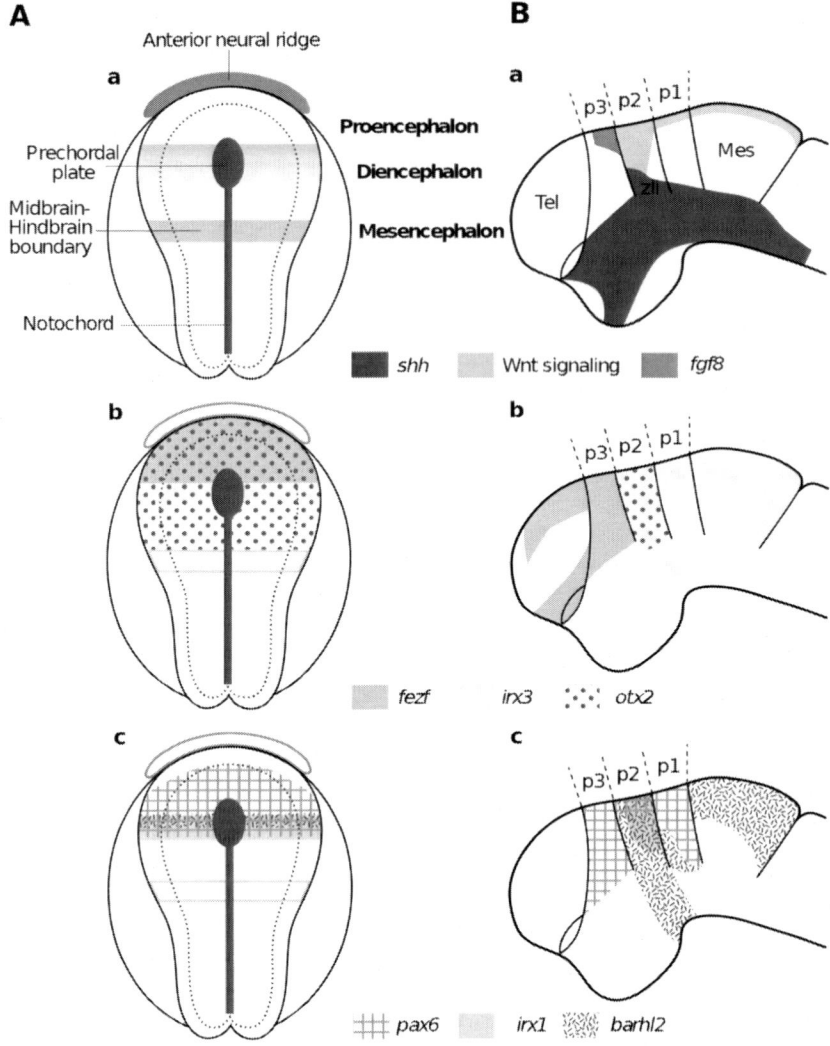

Figure 1. Schematic of *Xenopus* (*X.*) *laevis* neural plate and forebrain markers at st. 15 (A) and stage 30 (B). (A) Areas of expression are shown for (a) *shh* (red), *fgf8 (blue)*, Wnt signaling (ochre), (b) *fezf2* (violet), *irx3* (yellow) and *otx2* (violet dots) (c) *pax6* (orange squares), *barhl2 (*brown hash marks*)*, and *irx1/2* (pale green). The different parts of the developing neuroepithelium and the secreting structures are indicated including the Anterior neural ridge, the Prechordal plate, the Notochord and the Midbrain-Hindbrain boundary. p: prosomere; Tel: telencephalon; Mes: mesencephalon; *zli: zona limitans intrathalamica*.

The *zli* demarcates the boundary between pre-thalamus and thalamus and separates the posterior diencephalon (p1 and p2) from the anterior diencephalon (p3) (Figure 1; (Garcia-Lopez et al. 2004, Larsen, Zeltser, and Lumsden 2001, Shimamura et al. 1995)). This position corresponds approximately to the junction between the prechordal neuraxis overlying the prechordal mesendoderm, and the epichordal neuraxis overlying the notochord (Vieira et al. 2005) (Figure 1). The *zli* is defined by a dorsal ward continuation of *shh* expression into the alar plate of the diencephalon (Figure 1Ba). The *zli* dorsal progression within the alar plate starts at stage 24 in frog (Juraver-Geslin, Gomez-Skarmeta, and Durand 2014), HH12 to HH26 in chicken (Larsen, Zeltser, and Lumsden 2001, Zeltser, Larsen, and Lumsden 2001, Garcia-Lopez et al. 2004), E9 in mouse, and between the 12- and 15- somite stage in zebrafish (Scholpp et al. 2007). It is unique in brain regionalization because it represents the only neural area in which Shh, normally a DV patterning signal, regulates anterior-posterior (AP) regionalization (Ericson et al. 1996, Watanabe and Nakamura 2000).

Shh Expression Initiates and Demarcates *Zli* Development

During emergence of the neural plate, *shh* is expressed in the prechordal plate and the notochord. Shh from the notochord will activates both the Hedgehog signaling pathway and its own neural midline expression in the overlying neuroepithelial cells ((Placzek et al. 1990, Chamberlain et al. 2008); reviewed in (Placzek and Briscoe 2005)). Whereas in the posterior part of the neural tube, *shh* gene expression remains restricted to the floor plate, in its anterior part - the mesencephalon and prosencephalon - *shh* expression spreads from the floor plate to the basal plate (Figure 1B).

It is only at the boundary between p2 and p3 that a triangle-shaped of *shh* expression extends dorsalward (i.e., lateral) on both sides of the diencephalic walls (Figure 1; reviewed in (Placzek and Briscoe 2005)). Studies in chicken demonstrated that a sequential induction process, initiated by Shh secreted by floor plate cells, underlies the appearance of the *shh*-expressing line of cells corresponding to the *zli* (Zeltser 2005). This homeotic induction of *shh* occuring from one cell to its neighbor is not fully understood. Besides involving a specific Shh transcriptional regulatory network, the Shh maturation and secretion process (Section 5), together with changes at the level of the primary cilium that mediates Shh read-out (Section 6), could be involved. Of note because the *zli* is not formed through cell migration from the basal to the alar plate, the *zli* is considered an alar structure (Zeltser 2005).

Based on observations in frog, chicken, and mouse the accepted view is that a continuous source of Shh signals, provided *in vivo* by secretion of Shh initially from floor plate cells and then from the basal plate, is strictly necessary both for induction of *shh* expression within the *zli*, and for the correct segregation of *zli* cells from the thalamus (Scholpp et al. 2007, Vieira and Martinez 2006, Zeltser 2005, Juraver-Geslin, Gomez-Skarmeta, and Durand 2014, Garcia-Lopez et al. 2004). Of note in zebrafish, a small domain of *shh* expression at the position of the *zli* appears in *smoothened* (*smo*) mutants, in which signaling downstream of all Hedgehog family ligands is defective (Mathieu et al. 2002, Holzschuh, Hauptmann, and Driever 2003). Similarly, a population of "*zli* cells" develops in a quarter of zebrafish *cyclops* mutants, which lack a floor plate in the diencephalon (Sampath et al. 1998, Scholpp et al. 2007, Vieira and Martinez 2006, Zeltser 2005). Therefore, contrary to observations in other model organisms, the continuous presence of a Shh signal is not required for emergence of the *zli* in zebrafish.

The *Zli* Is a Tissue Compartment, Which Cells Segregates from Their Neighbors

Cells of a signaling center fulfill a localized patterning function. They need to be set as a coherent compartment with clear boundaries and lineage restriction enabling the signaling center to maintain its position relative to the surrounding tissue (Figdor and Stern 1993, Larsen, Zeltser, and Lumsden 2001, Garcia-Lopez et al. 2004, Martinez and Puelles 2000) (reviewed in (Kiecker and Lumsden 2005, Martinez-Ferre and Martinez 2012)). Analysis of chick diencephalon development showed that the caudal forebrain territory is not overtly segmented before appearance of the *zli* (Larsen, Zeltser, and Lumsden 2001). Emergence of the *zli*, however, correlates with acquisition of cell lineage-specific properties: at both the anterior and posterior limits of the *zli*, cell movements become restricted. Indeed *zli* formation is associated with acquisition of properties observed at tissue boundaries, specifically markers of immiscible interfaces such as chondroitin sulfate proteoglycans, laminin, weakly polysialylated neural cell adhesion molecules, and vimentin are detected within the *zli* territory and interkinetic nuclear migration movements are disrupted (Larsen, Zeltser, and Lumsden 2001, Garcia-Lopez et al. 2004). These studies established that the *zli* is a narrow compartment with cell lineage-restricted boundaries, and the first compartment to emerge in the forebrain. This compartment physically separates the diencephalon into the prethalamus anteriorly and the thalamus posteriorly (Larsen, Zeltser, and Lumsden 2001, Garcia-Lopez et al. 2004, Figdor and Stern 1993).

According to the prosomeric model the forebrain becomes subdivided into the so-called secondary prosencephalon (p4 to p6) that contains the telencephalon, optic vesicles, and hypothalamus, and the alar plate of the diencephalon that is divided into the p3, p2 and p1 (Figure 1)(Garcia-Lopez et al. 2004)(reviewed in (Martinez-Ferre and Martinez 2012, Puelles and Rubenstein 2003, Puelles 2018, Puelles et al. 2013)). Cell lineage analysis indicates that cells do not segregate between the secondary prosencephalon and the prethalamus (p3) (Larsen, Zeltser, and Lumsden 2001). Moreover, the competence of the prethalamus (p3) to respond to instructive factors such as Fgf8 is different from that of the other diencephalic prosomeres (p2 and p1) (Crossley, Martinez, and Martin 1996, Robertshaw et al. 2013). Therefore the *zli* separates territories with different adhesive properties and differential competence to respond to morphogens (Rubenstein et al. 1998, Larsen, Zeltser, and Lumsden 2001).

In conclusion, the *zli* is a compartment that expresses *shh*, situated in the alar plate of the caudal forebrain. Amongst the brain signaling centers it forms last. It is crucial for correct diencephalic patterning and introduces segmentation within the caudal forebrain. Whereas the *zli* is characterized by the same positioning, and similar gene expression patterns in all model organisms studied, observations in chicken suggest that its formation may vary between species. Studies in zebrafish, frog, chicken and mouse now allow proposing a model for the TFs and the inductive cues allowing formation of this unique signaling center.

PRE-PATTERNING OF THE *ZLI* TERRITORY: A ROLE FOR WNT CANONICAL SIGNALING

Receptors, ligands and modifiers of the Wnt pathway are expressed during caudal forebrain regionalization. Specifically, the expression of Wnt ligands Wnt3, Wnt3a, and Wnt8b that activate Wnt canonical signaling, marks the alar plate of p2 and the *zli* primordium (Figure 1) (Zeltser, Larsen, and Lumsden 2001, Martinez-Ferre et al. 2013, Juraver-Geslin, Gomez-Skarmeta, and Durand 2014). The key step in the transcriptional activation mediated by the canonical Wnt pathway is the nuclear increase of β-Catenin levels. This increase is mediated by a temporally-controlled inhibition of Glycogen Synthase Kinase 3β (GSK3β) activity that normally drives β-Catenin to 26S proteasome mediated degradation (Figure 2B)(reviewed in (Ramakrishnan et al. 2018, Cinnamon and Paroush 2008, Jennings and Ish-Horowicz 2008)). The Wnt canonical pathway is crucial for both patterning and proliferation of neural progenitors and is involved in Antero-Posterior (AP) patterning of the forebrain (Houart, Westerfield, and Wilson 1998) (reviewed in (Wilson and Houart 2004)). Observations in zebrafish, amphibian, and chicken have highlighted early and late roles for Wnt ligands in *zli* positioning, induction, and development (Mattes et al. 2012, Martinez-Ferre et al. 2013) (Figure 2B; 3).

In zebrafish, Wnt3 and Wnt3a have been shown to be required for survival of zli anlage cells. Decrease of canonical Wnt signaling using a Wnt signaling antagonist induces loss of the pre-zli territory. Conversely, the enhancement of Wnt signaling brought about by the GSK3β inhibitor BIO, leads to a larger zli revealed by a broader expression domain of shh (Mattes et al. 2012). Morpholino mediated depletion of both Wnt3 and Wnt3a leads to an increase of apoptosis and to a loss of the diencephalic organizer primordium. Interestingly, the effect of canonical Wnt signaling on the survival of zli anlage cells is restricted to a time window of 4 hours during somitogenesis. In embryos depleted for Wnt3 and Wnt3a the size of the prethalamic (fezf2), and thalamic (irx1b), markers expression domains are unaltered (Figures 1 and 3). Besides in Wnt3/3a depleted embryos the concomitant depletion of either Fezf2, or of Irx1b, activity rescues zli formation. These data suggest that both Fezf2 and Irx1b normally restrict the zli territory (Mattes et al. 2012). Therefore, whereas Wnt3/3a function is required for maintenance of the zli anlage, it is not for the maintenance of the prethalamus or thalamus territories.

Besides a requirement for a Wnt8b-mediated signal has been demonstrated to be necessary as a permissive step for the subsequent induction of *shh* expression and emergence of the *zli* in the diencephalic primordium (Martinez-Ferre et al. 2013). Gli3 is a Shh-regulated transcriptional repressor (Figure 2A) (Ruiz i Altaba, Nguyen, and Palma 2003, Jacob et al. 2003, Jessell 2000). During the neural plate patterning stages, *gli3* transcripts are detected within the alar neural plate and specifically in the alar diencephalon. In chick, this wide expression domain gets restricted to a narrow band of cells at the center of the *wnt8b* expression domain. The transverse alar stripe of *wnt8b* expression domain, which is now devoid of *gli3* expression, is the prospective *zli* anlage (Figure 3)(Hashimoto-Torii et al. 2003). Over-expression of *gli3* within the pre-*zli* territory inhibits *shh* induction, indicating that the local repression of *gli3* is necessary to allow *shh* homeotic induction within the pre-*zli* territory. Furthermore, the local downregulation of *gli3* within the future *zli* territory requires a Wnt-signal, which in chicken is mediated by the presence of Wnt8b. The entire domain that expresses *wnt8b* does not lose *gli3* expression demonstrating that whereas Wnt8b signal appears necessary to locally down-regulate *gli3* within the *zli* anlage, it is however, not sufficient.

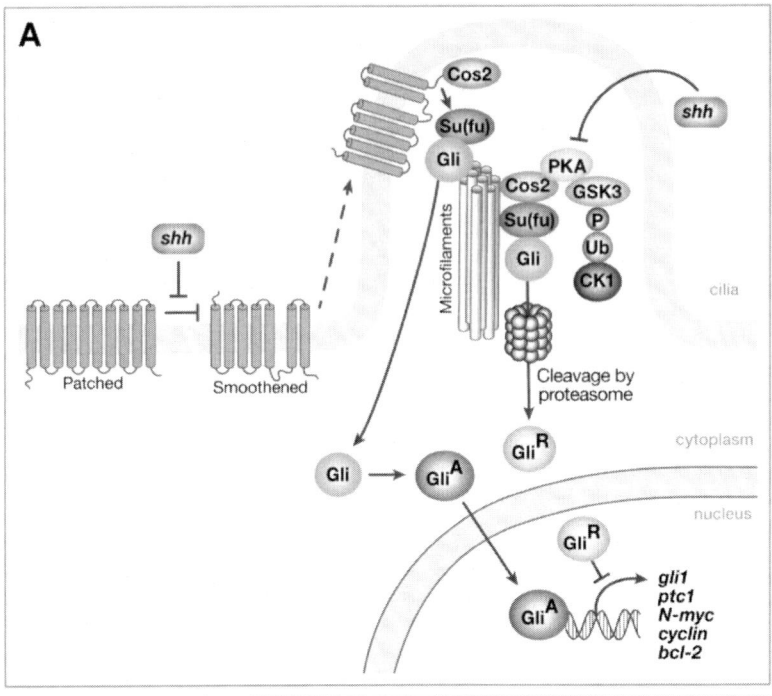
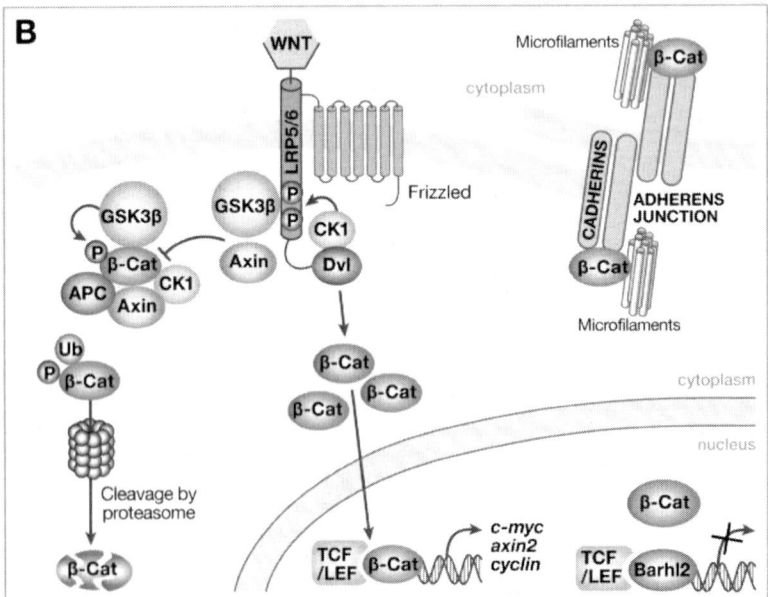

Figure 2. (Continued).

Figure 2. The Wnt canonical and the Sonic Hedgehog (Shh) pathways. (A) Schematic representation of the Shh pathway. In the absence of Shh, Patched (Ptc) inhibits Smoothened (Smo) and prevents its translocation to the cilium. In the absence of Shh the kinesin-like protein Costal-2 (*Cos2*) binds to the microtubules allowing the formation of a multiprotein complex with recruitment of the serine/threonine kinase Glycogen Synthase Kinase 3 (GSK3), Casein Kinase alpha 1 (CK1α), Protein Kinase A (PKA) and Suppressor of fused (Su(Fu)). Through a mechanism necessitating movements into, within, and out of the cilium, via intraflagellar transport the proteins complex phosphorylates the Gli proteins and targets their proteosomal degradation generating the repressor form of Gli, Gli^R. Truncated Gli forms (Gli^R) translocate to the nucleus where they repress target gene expression. Binding of Shh to Ptc initiates the pathway, alleviates repression of Smo, which is translocated to the cilium, and promotes cell survival. Cos2 binds to Smo and inhibits the phosphorylation of Gli proteins by releasing the different kinases. Full activation of the pathway is achieved by phosphorylation of the cytosolic C-tail of Smo by PKA and CK1. Full-length forms of Gli are translocated to the nucleus where they induce the expression of Shh target genes such as *gli-1*, *ptc*, N-*myc*, *bcl-2* and *cyclin D1*. (B) Schematic representation of the Wnt canonical pathway. In the absence of a Wnt signal, β-Catenin is phosphorylated and targeted for proteasome-mediated degradation by a protein complex containing GSK3β, CK1α, Adenomatous Polyposis Coli (APC) and Axin. Following binding of the Wnt ligand to the receptors Frizzled (Fz) and (Low-Density Lipoprotein Related Protein (LRP) 5/6, Dishevelled (Dvl) binds to Fz and recruits the destruction complex at the cytosolic tail of LRP5/6. LRP5/6 tail is then phosphorylated by GSK3β and bound by Axin. In this context, β-Catenin is stabilized and translocated to the nucleus where in association with TCF/LEF transcription factors it stimulates expression of the target genes *c-myc*, *axin2* and *cyclin D1*. In parallel, β-Catenin is present at the adherens junctions where it associates with cadherins and with the cytoskeleton allowing adhesion between neighbouring cells. In the presence of the TF Barhl2, β-Catenin cannot activate TCF/LEF mediated transcription.

Finally, once Shh is expressed in the *zli,* inhibition of the Wnt pathway does not have any effect on its maintenance (Martinez-Ferre et al. 2013). Besides Wnt8b, other Wnt ligands are present in the alar part of the p2 compartment including Wnt3, Wnt3a, Wnt4 and Wnt2b. Whereas in chick Wnt8b is the main ligand repressing *gli3* expression, other Wnts may fulfill a similar role in other species (Zeltser, Larsen, and Lumsden 2001, Martinez-Ferre et al. 2013, Juraver-Geslin, Gomez-Skarmeta, and Durand 2014).

In conclusion, Wnt-mediated signals are involved in maintenance of the *zli* anlage, and to generate the permissive conditions for the activation of Shh expression within the *zli*.

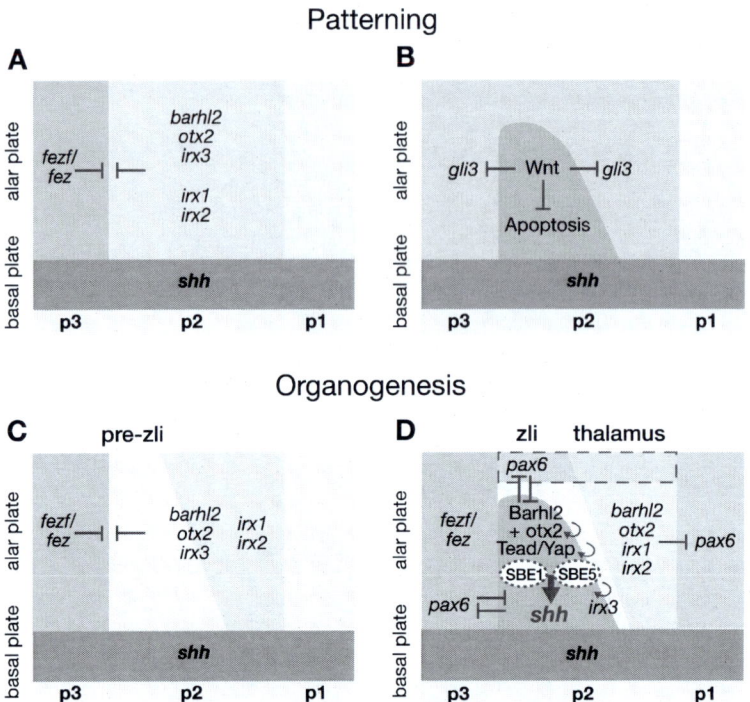

Figure 3. A model for building the *zli*. (A) At the neural plate stage, within the diencephalic primordium *otx2*, *barhl2* and *irx1*, *irx2*, *irx3* (pale green) are expressed. Cross-repressive interactions between Fezf/Fez and Irx homeoproteins participate to formation of the p2/p3 alar border. *shh* (red) is expressed in the basal plate. (B) Concomitantly, a Wnt-mediated signal promotes survival of *zli* anlage cells. Wnt canonical signal represses *gli3* expression in a narrow band of cells that correspond to the *zli* primordium. (C) At the onset of *zli* formation the *zli* anlage is characterized by the expression of *barhl2*, *otx2*, *irx3* (yellow), and the future thalamus by the expression of *barhl2*, *otx2*, *irx1/2/3* (green). The pre-zli primordium slowly becomes permissive to *shh* expression. (D) Shh secreted by basal plate cells initiates the spreading of *shh* expression into the pre-*zli* domain. Transcription factors (TFs) including Otx2, Barhl2, Yap–Tead, and possibly unidentified TFs (TF-X) are recruited to the enhancers SBE1 and SBE5 to initiate *shh* transcription. Within the pre-zli territory the homeotic induction process is transmitted sequentially from one cell to the next. Concomitantly *irx3*-expressing cells located in caudal p2 sort out from rostral thalamic cells (red arrows) under the inductive influence of secreted N-Shh. Shh signaling from the *zli* induces the repression of *pax6* within the mid-diencephalic furrow. Pax6 together with unidentified dorsal signals could in turn prevent *Shh* from being expressed beyond the *zli* in the future habenula (violet dashed rectangle). The thalamus (p2) and the prethalamus (p3) receive a combination of Shh signals coming from two orthogonally oriented signaling centers, the *zli* and the basal plate. The thalamus will separate in a rostral thalamus (yellow) and a caudal thalamus (green). Together, they establish a morphogenetic gradient of Shh protein that patterns the thalamus posterior to the *zli* and the prethalamus anterior to the *zli*. p: prosomere; *zli*: *zona limitans intrathalamica*.

LAYING THE GROUND FOR THE *ZLI:* OTX2 AND BARHL2 BINDING TO *CIS*-REGULATORY SEQUENCES CONFERS COMPETENCE FOR *ZLI* FORMATION

During the early steps of neural induction, an underlying pre-pattern emerges in the neural plate. The patterning cues influence the way in which neighboring cell populations respond differentially to similar morphogens (Kobayashi et al. 2002, Gomez-Skarmeta, Campuzano, and Modolell 2003, Robertshaw et al. 2013). Neural plate pre-patterning is partly encoded by the activity of TFs, which influence the neuroepithelial cell's response to inductive signals. Thereby analysis of TF expression dynamics in the anterior neural plate provides crucial information about the pre-patterning cues involved in emergence of the *zli* (Figure 1).

Two TFs are especially involved in allowing the future *zli* tissue to express *shh*. Orthodenticle homeobox (Otx) 1 and 2 are homeodomain-containing proteins involved in specification and regionalization of the forebrain (reviewed in (Acampora et al. 2000, Beby and Lamonerie 2013)). At gastrula and early neural stages *otx2* expression territory marks the anterior neural plate (Pannese et al. 1995). In both zebrafish and frog, *otx* expression decreases in the telencephalic territory during neurulation. At the onset of *zli* development *otx* expression is restricted to the p2, and midbrain, territories (Figure 1; 3) (Juraver-Geslin, Gomez-Skarmeta, and Durand 2014, Scholpp et al. 2007). The Bar-class homeodomain-containing (BarH) Barh1 and Barh2 are also homeodomain-containing TFs (reviewed in (Schuhmacher et al. 2011, Reig, Cabrejos, and Concha 2007, Juraver-Geslin and Durand 2015)). Transcripts encoding BarH-like (*barhl*) 2 are detected in the diencephalic primordium of amphibian (Offner et al. 2005, Juraver-Geslin et al. 2011), zebrafish (Staudt and Houart 2007), and mouse (Mo et al. 2004, Yao et al. 2016). Similar to *otx* genes, at the onset of *zli* development *barhl1* and *barhl2* expression are restricted to prosomere p2. While *barhl2* is expressed in the entire p2, *barhl1* expression is restricted to basal p2 (Figure 1; 3) (Patterson et al. 2000, Colombo et al. 2006).

In zebrafish, the lack of Otx1l/2 (the zebrafish ortholog of Otx2) function leads to absence of *zli* and thereby of *zli*-target territories. In embryos lacking Otx function, the thalamus is mis-specified prior to, and independently from, *zli* formation. The prethalamus and pretectum adjacent territories expand into the mis-specified territory and form a new interface. Similarly mouse embryos with reduced *otx1/2* transcripts display a lack of *shh* expression within the *zli* territory (Yao et al. 2016). Therefore, Otx1/2 presence is required to establish a competence area that allows *zli* formation (Hirata et al. 2006, Scholpp et al. 2007, Yao et al. 2016). In Otx-depleted zebrafish embryos, ectopic expression of Otx rescues formation of the *zli* however solely within the diencephalic territory, and anterior to the presumptive thalamus (Scholpp et al. 2007).

In *X. laevis* embryos depletion of Barhl2 abolishes development of the *zli* (Juraver-Geslin, Gomez-Skarmeta, and Durand 2014). Similarly, in mice deficient for the *barhl2* locus, *shh* expression within the *zli* is significantly reduced (Yao et al. 2016, Ding et al. 2016). *X. laevis* embryos depleted of Barhl2 resemble zebrafish embryos depleted of Otx1l/2: whereas most forebrain markers are unaltered, the embryos exhibit defects in *shh* expression in the *zli* and in the formation of the mid-diencephalic furrow (Ding et al. 2016, Juraver-Geslin, Gomez-Skarmeta, and Durand 2014). Of note, in zebrafish the loss of Otx2 generates a loss of *barhl2* expression whereas *otx2* expression is maintained in Barhl2-depleted *X. laevis*. Therefore, it is possible that Otx1l/2 proteins are involved in maintenance of *barhl2* expression in prosomere p2, and the loss of *barhl2* probably contributes to the *zli* defects observed in Otx-deficient zebrafish (Scholpp et al. 2007, Juraver-Geslin, Gomez-Skarmeta, and Durand 2014).

Barhl2 and Otx2 synergistic activities in *zli* formation are confirmed by a thorough analysis of the *cis*-regulatory-motifs controlling *shh* expression within the mouse *zli*. Two enhancers, SBE1 and SBE5, which specifically drive *shh* expression within the *zli* territory, have been identified. The enhancers activity is dependent on six position-independent motifs that are regulated by Otx2, Barhl2 and the TEA-domain family member 2 (Tead2), the key mediator of Hippo signaling, and its co-transcriptional activation partner Yap (Figure 3D). Yao et al. used large-scale genomic approaches associated with bioinformatic analytical tools, to characterize the 116-bp homology block, referred to as SBE1-like-enhancer. This enhancer is present in a scattered manner throughout the mouse and human genomes and its sequence is conserved from human to zebrafish. Using luciferase reporter assays, chromatin immunoprecipitation (ChIP), and transgenic mouse reporter assays, Yao *et al.* demonstrated that the six motifs, which they paired with known TFs, are both necessary and sufficient for full enhancer activity. Motifs 1 and 6 correspond to recognition sequences for Otx1/Otx2 and Barhl2 respectively. Combined action of both TFs resulted in a synergistic induction of reporters under motifs 1 and 6 control. Similarly, Tead2 and Yap recognize and contribute to *shh* expression through binding on motif 2. In mice, conditional loss of Yap1 demonstrated a selective reduction in *zli shh* expression. Tead2 and Yap activities on *zli* formation have not been assessed in other species. The second enhancer SBE5 contains a cluster of permuted motifs similar to those identified in the SBE1 enhancer but does not display any other overt sequence homology. SBE5 is located in the vicinity of the *shh* locus and works in a partly redundant manner with SBE1. In cell-based reporter and ChIP assays SBE5 performs equivalently to SBE1. Deletion of both enhancers in mouse entirely abolishes expression of *shh* within the *zli* (Yao et al. 2016).

Taken together, these studies reveal the presence of a "*zli* developmental cassette" that uses two main TFs, Otx2 and Barhl2, in combination, at least in mice, with the Tead2-Yap1 activation complex. These TFs through their interactions with conserved *cis*-regulatory motifs induce *shh* expression in a narrow band of cells within the caudal forebrain.

PART OF THE NETWORK INVOLVED IN ZLI FORMATION IS HIGHLY EVOLUTIONARILY CONSERVED

Chordate phylum conservation of the *cis*- and *trans*-regulatory landscape underlying *zli* development has been investigated. *Saccoglossus kowalevskii* (*S. kowalevskii*) is closest to the central basic reference animal at the root of the chordate phylogenetic tree (Pani et al. 2012, Lowe et al. 2003). In *S. kowalevskii*, the narrow band of cells at the proboscis-collar boundary is considered *zli*-like as it expresses hedgehog (*hh*). *S. kowalevskii barhl2* (*barH*), *otx*, and *irx* orthologs have been identified. Analysis of their expression patterns demonstrate that the orthologs of *barhl2*, *otx* and *irx* are expressed at the proboscis-collar boundary at the right time and place to perform their patterning function(s) (Pani et al. 2012, Lowe et al. 2003). Moreover Yao et al. identified a 1.1-kb region containing a *cis*-regulatory element containing the six motifs of the mouse SBE1 in *S. kowalevskii*: skSBE1. Expression of SBE1 in *S. kowalevskii*, or of skSBE1 in mice, demonstrates that skSBE1 is a functional ortholog of the mmSBE1 enhancer (Yao et al. 2016). In lamprey and all jawed vertebrates that display "*zli*-like" structures similar SBE1 *cis*-regulatory elements, intact or with a shuffled motif arrangement, were discovered. In contrast, similar SBE1-like motifs were not found in amphioxus (cephalochordate) or in ascidian (tunicate) that both lack an *hh*-expressing domain in the anterior brain.

In conclusion, these studies support the hypothesis that early chordates inherited an *hh cis*-regulatory-motif from a deuterostome ancestor that was subsequently lost in the invertebrate chordate lineages. A conserved *hh cis*-regulatory-motif (SBE1-like) was maintained in the vertebrate *shh* gene and used to activate its transcription in the *zli*, paving the way for the establishment of this brain-signaling center more than 500 millions years ago (Yao et al. 2016).

THE SONIC HEDGEHOG SEQUENTIAL INDUCTION PROCESS: A SECRETED SIGNAL READ OUT THROUGH A PRIMARY CILIUM

During *zli* formation the N-terminal part of Shh secreted by basal plate cells is released into the ventricular lumen and N-Shh activates its own expression in neighboring cells (Zeltser 2005). Whereas pre-*zli* cells respond to secreted Shh, neither its neighboring prethalamic, or thalamic, cells will.

Wnt signaling, the absence of *gli3*, and the presence of specific TFs are all required for this homeotic induction process to happen. However molecular cues controlling sequential induction and the differential response to Shh are not fully deciphered (also see Section 10). Modifications in Shh maturation and secretion, and/or in the primary cilium and/or in the read-out of Shh could participate in this unique process.

Shh Maturation and Secretion

For Shh proteins to be secreted into the ventricular lumen an autocatalytic internal cleavage associated with the addition of lipid molecules, specifically cholesterol and palmitic acid moieties is necessary (reviewed in (Ramsbottom and Pownall 2016)). Shh cleavage confers a hydrophobic character to the molecule that promotes its association with the cell membrane. Indeed Shh proteins truncated at the site of internal cleavage diffused more widely (Porter et al. 1996). The internal cleavage takes place in the endoplasmic reticulum and produces a 25 kD carboxy-terminal domain (C-Shh) and a 20 kD amino-terminal domain (N-Shh) (Porter et al. 1995, Bumcrot, Takada, and McMahon 1995, Chen et al. 2011). The C-Shh part recruits a cholesterol molecule that is further attached to the C-terminus of N-Shh (Porter, Young, and Beachy 1996, Chen et al. 2011). The binding of cholesterol to N-Shh limits the range of Shh signaling and is consequently crucial for its function. In the mouse limb bud, N-Shh lacking cholesterol has an extended signaling range (Li et al. 2006). Besides cholesterol, the attachment of a palmitic acid group on the N-terminal part of N-Shh has proved necessary for its activity (Pepinsky et al. 1998). An acyltransferase named Skinny Hedgehog (SKI) in *Drosophila* and mouse and Hedgehog acyltransferase (HHAT) in humans mediates the transfer of the palmitic acid group (Buglino and Resh 2008, Chamoun et al. 2001, Hardy and Resh 2012). In mice, the lack of N-Shh palmitoylation decreases Shh signaling in both limb buds (Chen et al. 2004) and ventral forebrain (Kohtz et al. 2001).

N-Shh is strongly hydrophobic and different mechanisms have been described for the release of processed N-Shh (pN-Shh) into the extracellular space. In *fly*, the transmembrane protein Dispatched-1 (Disp1) and the secreted protein Scube2 interact with pN-Shh. Disp1 and Scube2 interact with different parts of the cholesterol attached to N-Shh and thereby promote the release of pN-Shh from the cell surface (Burke et al. 1999, Creanga et al. 2012, Tukachinsky et al. 2012). Other evidence indicates that a lipoprotein complex activity mediates pN-Shh secretion. The strong hydrophobicity of the pN-Shh molecule is supposedly decreased via the phospholipid monolayer of the lipoproteins that embraces the lipids present in the pN-Shh and leave the hydrophilic areas outside (Eugster et al. 2007, Panakova et al. 2005). In *Caenorhabditis elegans* (*C. elegans*) (Liegeois et al. 2006) and *Drosophila* (Callejo et al. 2011), an exosome-mediated release has been described. In vertebrates pN-Shh can also be released in a vesicular form (Vyas et al. 2014).

The Shh Signal Is Read through a Primary Cilium

Shh signaling in vertebrates requires the presence of primary cilia: small, membrane-sheathed cell protrusions that occur on almost all cells during development and adulthood (Huangfu et al. 2003)(reviewed in (Wheway, Nazlamova, and Hancock 2018, Bangs and Anderson 2017, Park, Jang, and Lee 2019). Whereas motile cilia bent via dynein motors that promote their rotational or beating movement, primary cilia are immotile. Primary cilia are highly specialized sensory organelles: their membranes are studded with a variety of different chemo- and mechano-receptors (Hilgendorf, Johnson, and Jackson 2016). Sensory primary cilium signal transduction necessitates intraflagellar transport (IFT): a motor protein-driven transport along axonemal microtubules. Indeed the Hedgehog signal transduction machinery including Patched (Ptc), a twelve-transmembrane domain receptor, Smo, a G-protein coupled receptor, and the zinc finger containing Gli family of TFs are all moving into, within, and out of the cilium, via IFT depending on Shh pathway activation status (Figure 2A) (Bai, Stephen, and Joyner 2004); (reviewed in (Briscoe 2009, Fuccillo, Joyner, and Fishell 2006, Ramsbottom and Pownall 2016, Mukhopadhyay and Rohatgi 2014)). Shh binding to Ptc leads to the relief of Smo inhibition at the cilium, which promotes an intracellular transcriptional response that depends on the Gli TFs. A balance between the production of activating forms (Gli^A) and repressor forms of Gli (Gli^R) is at the core of the Shh response. In mammals, there are three Gli proteins (Gli 1 to 3). When Shh signaling is off, Gli2 and Gli3 are phosphorylated, targeted to the proteasome, and then cleaved. The cleaved forms correspond to repressive forms (Gli^R) and $Gli3^R$ is the strongest transcriptional repressor of the Shh pathway. When Shh ligand is present at the cilium, the cleavage of Gli proteins is abolished and Gli proteins transcriptionally activates the Shh target genes (Gli^A) (Figure 2A)(reviewed in (Briscoe 2009)).

During embryogenesis, neuroepithelial cells in mouse (Huangfu et al. 2003, Willaredt et al. 2008), chicken (Cruz et al. 2010), *X. laevis* (Chung et al. 2012), and zebrafish (Kramer-Zucker et al. 2005) carry primary cilia on their apical surface in contact with the ventricular fluid. DV patterning of the neural tube relies on cilia-mediated Shh signal transduction (Huangfu et al. 2003). In the mouse forebrain, cilia have been identified on cells in contact with the ventricles (Christ et al. 2012) where they allow $Gli3^R$ formation and DV patterning (Besse et al. 2011, Laclef et al. 2015). In the diencephalon cilia have been detected in cells corresponding to the region of the *zli* in both *X. laevis* (Hagenlocher et al. 2013) and mice (Andreu-Cervera et al. 2019). Noteworthy, primary cilia observed on cells within the *shh*-expressing brain signaling centers including the floor plate, and the *zli* are elongated compared to surrounding cilia present on non-*shh*-expressing cells (Hagenlocher et al. 2013, Cruz et al. 2010, Chung et al. 2012). As longer cilia seem to correlate with a reduced activation of the pathway this could indicate a way of differentially adapting signaling pathway activation (Cruz et al. 2010) (Section 12).

In conclusion, in the context of *zli* development mechanisms mediating Shh secretion and controlling its diffusion rate are currently investigated. N-Shh strong hydrophobic character suggests that it diffuses very slowly and tends to stay close to the membrane surface at least partly explaining the sequential induction process observed during *zli* development. Moreover, a primary cilium is present on diencephalic progenitor cells in vertebrates. The primary cilium is structurally different on pre-*zli* cells versus both prethalamic and thalamic cells. Whereas little is known on the biological consequences of such differences the phenotypic analysis of ciliary mutants argue that the net read out of Shh signal within *zli* cells is unique.

BUILDING OF A COMPARTMENT: A ROLE FOR THE IROQUOIS COMPLEX IN REFINING THE ZLI BORDERS

The Iroquois (*irx*) genes encode for homeodomain-containing TFs, highly conserved from *Drosophila* to mammals (reviewed in (Gomez-Skarmeta and Modolell 2002, Cavodeassi, Modolell, and Gomez-Skarmeta 2001)). Most vertebrates contain six *irx* genes grouped in two paralog clusters of three genes each (Peters et al. 2000, de la Calle-Mustienes, Modolell, and Gomez-Skarmeta 2002). The *irxA* cluster contains *irx1*, *irx2* and *irx4*, while the *irxB* cluster corresponds to *irx3*, *irx5* and *irx6*, and in zebrafish, *irx7*. The Irx proteins participate in defining territories and in specifying cell identity. *irx* gene activities in boundary formation have been previously described (reviewed in (Gomez-Skarmeta and Modolell 2002)).

In all species studied, *irx1, irx2,* and *irx3* are co-expressed in the anterior neural plate during the early stages of neural patterning where they mark the future p2 (Figures 1 and 3; (Rodriguez-Seguel, Alarcon, and Gomez-Skarmeta 2009, Offner et al. 2005, Lecaudey et al. 2004)). In zebrafish, *irx3b* is strongly expressed in the developing *zli* together with *irx5a, irx5b* and *irx7*. The *irx3b/5/7* expression pattern subdivides the *zli* into distinct DV domains. In contrast, *irx1a, irx1b, and irx2a* expression domains abut the *zli* posteriorly (Lecaudey et al. 2004). *In X. laevis* analysis of *irx* genes expression during diencephalic primordium development reveals a temporally dynamic process (Juraver-Geslin, Gomez-Skarmeta, and Durand 2014, Rodriguez-Seguel, Alarcon, and Gomez-Skarmeta 2009). At neurula stages *irx1, 2* and *3* are co-expressed in the p2 domain together with *barhl2* and *otx2*. At the onset of *zli* emergence the rostral part of p2 - the future *zli*

- only expresses *irx3,* whereas the caudal p2 - future thalamus - expresses *irx1, 2* and *3*. At stage 35 when the *zli* has emerged, the rostral p2 expresses *shh* and *irx3,* whereas the caudal p2 has lost irx3 expression; it solely expresses *irx1* and *irx2* (Figures 1 and 3) (Juraver-Geslin, Gomez-Skarmeta, and Durand 2014, Rodriguez-Seguel, Alarcon, and Gomez-Skarmeta 2009).

Observations in zebrafish and amphibian promote the idea that Irx factors participate in acquisition of *zli* compartment identity (Scholpp et al. 2007, Juraver-Geslin, Gomez-Skarmeta, and Durand 2014). In zebrafish, *irx1b* function is dispensable for *zli* development; however, depletion of the *irx1* orthologs *irx1b* and *irx7* induces a posterior shift of the *zli* caudal border and an expansion of *shh* expression domain at the expense of the thalamic field (Hirata et al. 2006, Scholpp et al. 2007). In agreement with the observations in zebrafish, functional experiments in amphibian argue that the ratio of Irx3 to Irx1/2 is critical for both *zli* specification and establishment of its posterior boundary. Indeed, overexpression of *irx3,* or depletion of *irx1/2,* within the *barhl2/otx2*-expressing area promotes the acquisition of a *zli* fate at the expense of a thalamic fate.

Observations in chick embryos are divergent from those described in *zebrafish* and *X. laevis*. Chicken *irx3* is not expressed within the *zli* but posterior to it, and appears to have a repressive function on *zli* formation (Kiecker and Lumsden 2004). Misexpression experiments in the caudal forebrain through *in ovo* electroporation approaches, indicate that Irx3 together with another TF Pax6 (Section 10.3) alters the competence of caudal forebrain cells to respond to both Fgf8 and Shh. Irx3 ectopic expression in the prethalamic anlage induces thalamic markers and represses prethalamic markers. The authors suggest that Irx3 participates in establishment of the differential cellular competence to respond to Shh signaling from both side of the *zli* (Robertshaw et al. 2013).

In conclusion, the Iroquois TFs participate in establishment of the anterior and posterior borders of the *zli*, and in acquisition of its cell segregation properties. Irx activities may vary depending on species.

MAINTAINING A COMPARTMENT: SIGNALING PATHWAYS AND TFS THAT MODULATE ADHESIVE PROPERTIES

Specific cellular features characterize compartments: (i) compartment cells proliferate slowly and (ii) cells of flanking compartments separate along boundaries. Differences in the adhesive properties (affinity) of their cell surfaces mediate their segregation behavior. At the boundary, cells deposit an extracellular matrix that acts as a mechanical barrier separating different cell populations. Establishment of adhesive differences constitutes the first step of lineage restriction, whereas establishment of a mechanical fence further stabilize compartments (reviewed in (Kiecker and Lumsden 2005, Dahmann, Oates, and Brand 2011)).

Canonical Wnt Signaling in Diencephalic Cell Segregation

As previously mentioned Wnt canonical ligands are detected in the alar plate of prosomere p2 (Figure 1). β-Catenin is the main effector of the canonical Wnt pathway and mediates the interactions between the intracellular cytoskeleton and the Cadherins, a group of cell-cell adhesion proteins thought to be important in the formation of compartment boundaries (Figure 2B). Members of the Cadherin superfamily members are expressed differentially in the forebrain subdivisions and participate in the adhesive properties of the different histogenic fields (reviewed in (Redies 2000, Puelles 2007, Peukert et al. 2011)). In the zebrafish thalamus, β-Catenin regulates protocadherin 10b expression (Pcdh10b -, formerly known as OL-protocadherin) (Peukert et al. 2011)). *pcdh10b* plays an important role in thalamic field segregation. Alteration of *pcdh10b* expression within the thalamus territory provokes the intermingling of thalamic cells with its neighbors, mostly with cells of the pretectum. Wnt driven stabilization and nuclear translocation of β-Catenin induces a broadening of the expression domain of *pcdh10b,* whereas Wnt canonical signaling inhibition decreases *pcdh10b* expression. Therefore, Wnt canonical pathway contributes to the acquisition of the adhesive differences underlying differential cell segregation behaviors within the prosomere p2. Whether Wnt canonical signaling participates in separation of the *zli* field from the prethalamus and/or thalamus, and which cell surfaces adhesion molecules, including Cadherins, contributes to this behavior has not been deciphered.

The Notch/Delta Pathway and Separation of the *Zli* and Thalamic Fields

Notch signaling mediates lateral inhibition in embryonic tissues and during neural development (reviewed in (Sjoqvist and Andersson 2019)). *radical fringe (rfng)* and *lunatic fringe (lfng)* are two glycosyltransferases involved in regulating Notch signaling. Both have been involved in the establishment of boundaries. In zebrafish, activation of the Notch pathway directs cells to the rhombomere boundaries, whereas inhibition of Notch activity excludes cells from boundaries (Cheng et al. 2004). In avian embryos, a triangular shape forebrain domain that is depleted of lfng expression characterizes the pre-*zli*. Electroporation experiments revealed that an ectopic expression of *lfng* in this pre-*zli* compartment results in sorting of the cells into the *lfng*-expressing border regions. The authors concluded that *lfng* contributes to the acquisition of *zli* compartment properties (Tossell et al. 2011, Zeltser, Larsen, and Lumsden 2001). Of note, this pre-*zli* domain is thought to collapse during development to form the final *zli* (Zeltser, Larsen, and Lumsden 2001). However, the mechanism allowing this allosteric growth are not yet explained, and are not attributable to either cellular movements in the epithelium (Garcia-Lopez et al. 2004) or to cell death (Zeltser, Larsen, and Lumsden 2001) . Of note the "compartment" quality of the forebrain domain depleted of lfng expression has been questioned (Puelles 2018).

Barhl2 and Irx TFs Participate in the Acquisition of *Zli* Compartment Properties

β-Catenin, the main effector of the canonical Wnt pathway mediates neuroepithelial cell proliferative response through transcriptional regulation of *myc* and *cyclin-D1* (Figure 2B) (reviewed in (Grigoryan et al. 2008, van Amerongen and Nusse 2009, Noelanders and Vleminckx 2016)). In chick and mouse analysis of caudal forebrain proliferation kinetics reveals that *zli* cells divide slowly relative to cells in their neighboring territories (Baek et al. 2006, Martinez and Puelles 2000). As described above, Barhl2 mediates *shh* transcriptional activation within the *zli*. Moreover, Barhl2 limits activation of the Wnt canonical pathway within the prosomere p2, thereby acting as a brake on neuroepithelial cell proliferation and contributing to maintenance of diencephalic neuroepithelial architecture (Juraver-Geslin et al. 2011). Recent studies demonstrate that Barhl2 prevents β-Catenin ability to transformed Tcf in a transcriptional activator (Sena et al. 2019) (Figure 2B). In fly imaginal discs reducing *irx* genes activity accelerates the G1 to S transition and promotes cell proliferation. Conversely irx increased expression induces cell-cycle arrest a process that participate to imaginal disc size determination (Barrios et al. 2015). Moreover in the *Drosophila* eye the Irx proteins participate in the establishment of distinct cell affinities in dorsal versus ventral cells (Villa-Cuesta, Gonzalez-Perez, and Modolell 2007). In zebrafish, *iro7* is required for the proper positioning of the prospective r4/r5 hindbrain rhombomeric boundary (Lecaudey et al. 2004, Jimenez-Guri and Pujades 2011). Taken together, these observations indicate that Barhl2 along with the Irx TFs could participate in reducing *zli* cell proliferation as well as enabling adhesion and facilitating acquisition of *zli* compartment properties (reviewed in (Juraver-Geslin and Durand 2015)).

In conclusion, Wnt, Notch, and Shh signaling pathways together with Barhl2 and Irx TFs coordinately participate in driving *zli* cells behavior. However, the molecular processes slowing *zli* cell cycle rate, and driving the segregation of *zli* and thalamic cells, still have secrets to reveal.

PROGRAMMING OF A "ZLI-LIKE" STRUCTURE IN X. LAEVIS DORSAL BLASTOCOEL EXPLANTS

In amphibian, cells from the roof of the blastocoel, referred to as Animal Cap (AC), can be isolated and programmed to generate "organoids" through either manipulation of gene expression by microinjection of mRNA, DNA and antisense morpholino oligomers (MOs), and/or induction by secreted factors. An investigation of the signaling network underlying acquisition of *zli* cellular identity was performed using such explants (Durand 2016, Juraver-Geslin, Gomez-Skarmeta, and Durand 2014).

AC explants co-expressing *barhl2*, *otx2* and *irx3* and exposed to a continuous source of Shh signal acquire a *zli*-like identity. They express *shh* and the shape of the *shh*-expressing domain observed in the programmed explants—broad where in contact with the source of Shh signal and pointed at its other extremity—is consistent with a sequential *shh* induction process occurring from one cell to the next. Concomitant in time with acquisition of a Shh-expression program, the *zli*-like cells acquire cell segregation properties that depend on the presence or absence of a Shh signal: thereby mimicking their *in vivo* segregation behaviors. When grafted into a developing neural plate and continuously exposed to a Shh signal, neuroepithelial cells co-expressing *barhl2*, *otx2,* and *irx3* form *in vivo* an ectopic *zli* (Durand 2016, Juraver-Geslin, Gomez-Skarmeta, and Durand 2014).

These approaches confirmed that the signaling network which governs the competence and behavior of *zli* cells can be reconstructed in an *ex vivo* system. They also confirm cell-autonomous activities for Otx2 and Barhl2 (Durand 2016, Juraver-Geslin, Gomez-Skarmeta, and Durand 2014). They also suggest that Shh secreted by *zli* cells participates to the recruitment of *irx3* expressing cells from the thalamus into the *zli*, thereby contributing to establish a compartment. Whether *irx3* expressing cells behave similarly *in vivo* has not been demonstrated. However, such recruitment would participate in giving the *zli* its triangular shape, wide at the bottom and narrow at its top, a morphology observed in most vertebrate species (Figures

1Bb and 3). Finally, these data also reveal that the efficiency of *shh* induction—i.e., the average size of the *shh*-expressing area—is increased in the presence of a thalamus-like explant (Durand 2016, Juraver-Geslin, Gomez-Skarmeta, and Durand 2014).

Whether the thalamus territory through a cell non-autonomous mechanism facilitates the induction of *shh* and contributes to the dorsal progression of *shh* expression is not known. However, these "organoids" should help in investigating this question. They should also help in identifying the transcriptional changes initiated by the co-expression of *otx2*, *barhl2*, and *irx* genes in the presence or absence of Shh, thereby revealing molecular cues driving segregation of *zli* and thalamic cells.

ZLI POSITIONING WITHIN THE FOREBRAIN AND PRE-PATTERNING OF THE THALAMUS

A Role for the p2/p3 Border

Using grafting experiments in chicken embryos, Vieira et al. showed that an ectopic border between neural tissue from a prechordal (defined as being *six3*-positive) and epichordal (defined as being *irx3*-positive) origin is sufficient to induce an ectopic *zli* (Vieira et al. 2005). In Xenopus embryos, analysis of *irx*, *barhl2* and *shh* expression patterns during neurulation support the idea that the *zli* forms close to the interface between the prechordal neuraxis, induced by prechordal plate mesoderm, and the epichordal neuraxis, induced by the chordamesoderm (Figure 1B)((Offner et al. 2005, Rodriguez-Seguel, Alarcon, and Gomez-Skarmeta 2009, Larsen, Zeltser, and Lumsden 2001, Shimamura et al. 1995)).

The Fez/Fezl TFs and the *Zli* Anterior Border

In all investigated species, the *zli* anterior border emerges at the interface between the expression domains of the FEZ family zinc finger 2 (*fezf2*), which marks the alar plate of p3 rostral to the *zli*, and *irx3* that marks p2 (Figure 1) (Offner et al. 2005, Shimamura et al. 1995, Zeltser 2005, Martinez-Ferre et al. 2013, Vieira and Martinez 2006).

The *fez-like (fezl)* genes are highly evolutionarily conserved. During early segmentation stages *fezl* genes are expressed in the presumptive prethalamus but are not detected within the *zli* territory (Figures 1A, Be and 3). Zebrafish and mouse functional analysis indicate a role for Fez and Fezl in *zli* formation however the results are somewhat controversial. E12.5 mouse embryos mutated for both *fez* and *fezl* genes display prethalamus and diencephalic developmental defects including a loss of *shh* expression in the *zli*, an extended *wnt3a* expression domain within the p2 alar plate and an ectopic expression of *pax6* in the mid-diencephalic furrow. In contrast, the knock down of *fezl* in zebrafish results in an anterior expansion of the *zli*, associated with an expansion of *irx3a*. This phenotypic difference in loss of function phenotypes remains unexplained. It could either be species-dependent or due to a difference in levels of FEZL expression. Indeed, whereas in the zebrafish *fezl* knockdown, the *fez locus* is still intact, the *fez/fezl* double mutant in mice would reflect true null conditions. Conversely, in both species the over-expression of *fezl*, or *fez*, abolishes expression of *shh* in the *zli*. In zebrafish over-expressing *fezl* an expansion of prethalamus and hypothalamus territories at the expense of the *zli* and posterior forebrain and/or mid-brain domains, is observed. Despite these differences in all studied species the *fezl/fez* gene activity, and specifically *fezf2* activity, is crucial for correct establishment of the p2/p3 boundary that always co-localizes with the *zli* anterior boundary (Figure 1) (Jeong et al. 2007, Hirata et al. 2006).

The midbrain–hindbrain boundary, which is a secondary brain-signaling center, develops at and is induced by the interface of *otx2* and *gbx2* expression domains. By analogy it has been suggested, but not tested, that the *fezf2* and *irx3* expression domains interface is the inductive cue at the origin of *zli* formation (reviewed in (Martinez-Ferre and Martinez 2012, Scholpp and Lumsden 2010, Epstein 2012, Hagemann and Scholpp 2012, Kiecker and Lumsden 2005)). Indeed, the p2/p3 boundary localizes at the border between *fezf2* and *irx3* expression domains but also that of *fezf2* and *otx2*, and that of *fezf2* and *barhl2* (Figures 1 and 3). In amphibian, neither over-expression of *irx3*, nor down-regulation of *barhl2*, affects formation of the p2/p3 border (Juraver-Geslin, Gomez-Skarmeta, and Durand 2014, Scholpp et al. 2007). Similarly, in zebrafish, the loss of *otx* does not affect *zli* anterior border location (Scholpp et al. 2007). It is possible that the p2/p3 border is set up at the interface between the Fezf/Fez TFs and redundant activities of Otx2, Barhl2 and/or Irx TFs. Alternatively, and supported by data obtained in frog explants and mice, whereas the p2/p3 limit appears to delimitate the *zli* anterior border, the co-expression of *barhl2*, *otx2* and *irx3* could be the initiating cue in *zli* development. This observation is important, as it indicates that the *zli* developmental mode may be different from that of the midbrain–hindbrain boundary.

Pax6 TF and the *Zli* Dorsal Limit

The paired box gene 6 (Pax6) TF is expressed in the diencephalic territory from neurulation onwards. At the neurula stage *pax6* expression extends from the retina fields to the limit between the pretectum (p1) and the midbrain. At the onset of *zli* development *pax6* expression is excluded from the mid-diencephalic furrow and restricted to the epithalamus that produces the habenula (Figure 1 and 3). Mice deficient for the *pax6* locus, so-called Small-eye mice, exhibit defects in their diencephalic morphology from E10.5 onwards including a diffuse diencephalic/midbrain boundary and a decrease of diencephalic progenitor proliferation rate (Caballero et al. 2014). Moreover the epithalamus is missing and appears to be replaced by a thalamic domain that is shifted dorsally (Warren and Price 1997, Chatterjee et al. 2014). These defects have been partly associated with alterations in Shh signaling coming from the *zli*. In both mice and zebrafish, deletion of Pax6 generates a precocious and expanded *zli* territory. Conversely Pax6 overexpression in zebrafish prevents *zli* development. In Pax6 mouse mutants the attenuation of Shh signaling restores development of the epithalamus, indicating that Shh acts downstream of Pax6 in partitioning the p2 domain into the epithalamus and thalamus (Chatterjee et al. 2014).

Results in chick appear slightly different. As previously mentioned, *pax6* alone or together with *irx3* changes the cellular response to Shh thereby contributing to establishment of the caudal and rostral thalamus. In this model *irx3* is necessary for rostral thalamus determination whereas *pax6* together with *irx3* are determinant in acquisition of a caudal thalamus identity (Robertshaw et al. 2013). Congruent with this observation, within the chick mid-diencephalic furrow *shh* signaling from the *zli* represses *pax6* (Kiecker and Lumsden 2004). An analysis of Pax6 mutant mouse chimeras, suggest that Pax6 cell autonomously blocks *shh* expression in cells surrounding the *zli*. Using Immunoprecipitation experiments and luciferase assays the authors claimed that Pax6 binds the promoter of *shh* and directly represses it (Caballero et al. 2014). Taken together these observations are difficult to reconcile. Analysis of the SBE1-like-enhancer sequence that drives mouse *shh* expression in the *zli* does not carry Pax6 binding motif (Yao et al. 2016). In Xenopus tadpole, at the onset of *zli* formation, *pax6* is mostly present in the epithalamus and pretectum but not in the rostral thalamus and not flanking the pre-*zli* territory (Juraver-Geslin, Gomez-Skarmeta, and Durand 2014). Grafting experiments in chicken demonstrate that grafts of dorsal diencephalic tissue inhibit *zli* propagation, indicating that the progression of *shh* expression is limited dorsally by inhibitory factors (Zeltser 2005) (Figure 1A,Bf). In agreement with Chatterjee et al. we favor the hypothesis that Pax6 together with unidentified dorsal signals limits *zli* dorsal progression and participate in partitioning the dorsal prosomere p2 in caudal thalamus and epithalamus (Chatterjee et al. 2014).

In conclusion, the molecular signals initiating *zli* positioning remain partially unknown. The TFs that participate to *zli* emergence also contribute to pre-patterning of the thalamus and modify the cellular competence to respond to Shh. In this regard, the determination of signals controlling *otx*, *barhl2*, *pax6* and *irx* neural plate expression should provide important information on diencephalic pre-patterning cues.

FOREBRAIN MALFORMATIONS AND THE *ZLI* DEVELOPMENTAL NETWORK

Genetic Disorders Associated to Defects in Diencephalic Early Development

The TFs participating to *zli* formation: *otx*, *barhl2*, *irxs*, are all involved in thalamic primordium induction, growth, patterning, and organogenesis. Depletion of *otx* genes prevents forebrain development (reviewed in (Acampora et al. 2000)). In both frog and mouse, *barhl2* targeted-depletion generates thalamus developmental defects (Juraver-Geslin, Gomez-Skarmeta, and Durand 2014, Ding et al. 2016). In mice deficient for the *barhl2* locus, p2 thalamic progenitors acquire a pretectal fate and there is an absence of thalamo-cortical axon projections (Ding et al. 2016). Similarly, *irx* gene depletion in frog, chicken, and zebrafish disrupts thalamus development (Rodriguez-Seguel, Alarcon, and Gomez-Skarmeta 2009, Robertshaw et al. 2013, Juraver-Geslin, Gomez-Skarmeta, and Durand 2014, Scholpp and Lumsden 2010, Llinas 2003, Martinez-Ferre and Martinez 2012, Chatterjee and Li 2012).

In humans and other model organisms some genetic disorders including holoprosencephaly (HPE) or the ciliopathies, are characterized or accompanied by brain malformations including in some cases thalamic and thalamo-cortical developmental defects (reviewed in (Marcorelles and Laquerriere 2010, Willaredt, Tasouri, and Tucker 2013)). Besides, abnormal neuronal activity within the thalamus has been associated with neuropsychiatric disorders such as obsessive-compulsive disorder and attention deficit hyperactivity disorder (reviewed in (Xavier et al. 2016,

Guerrini and Dobyns 2014, Hu, Chahrour, and Walsh 2014, Shepherd 2013, Llinas 2003)).

Early Neural Plate Patterning and Midline Formation at the Core of HPE Etiology

The most prominent human pathological condition in which development of the thalamus is affected is HPE, defined as a defect in the formation of midline structures of the forebrain and face (Muenke and Beachy 2000, Fernandes and Hebert 2008). HPE occurrence is of 1 case in about 16,000 live births but approximately 1/250 foetus are thought to be affected by HPE which makes it the most common forebrain defect in humans. HPE presents itself as variable degrees of fusion between the left and right halves of the cerebral hemispheres, basal ganglia and, in rare cases, also the thalamus (Marcorelles and Laquerriere 2010, Hahn and Barnes 2010). Rather than originating from a fusion event, HPE arises from a failure to separate the two halves of the forebrain along the midline. Current studies focus on correct separation between the cerebral hemispheres and little is known on the thalamic contribution to the etiology of HPE (reviewed in (Munke 1989)).

Mutations have been identified in HPE human patients and animal models of HPE. Importantly all of them are found in genes known to play important roles in major brain developmental signaling pathways. Whereas defects in the Shh pathway are the most frequent cause of HPE, genetic screening of HPE patients and studies in animal models also involved the Fgf, Nodal and Notch pathways as major contributors to HPE etiology. Moreover most of these mutations are in genes that participate in early neural plate patterning and midline formation, strengthening the hypothesis that developmental defects at the origin of these malformations take place early during neural development (Nanni et al. 1999, Gongal, French, and Waskiewicz 2011, Petryk, Graf, and Marcucio 2015, Dubourg et al. 2016, Dupe et al. 2011, Mercier et al. 2013).

Shh Signaling Defects Contribution

As described above, the Shh pathway plays a seminal role in diencephalon development, and relies on primary cilia for its signal transduction. It is thus not surprising that mutations in certain cilia-related genes affect diencephalon morphology and induce HPE or HPE-like phenotypes. Deletions in the anterograde intraflagellar transport (IFT) 172, mouse gene down-regulates forebrain Shh signaling. Consequently, the mutant diencephalon is severely reduced and embryos exhibit lobar or semilobar HPE (Gorivodsky et al. 2009). Diencephalon development is also impacted in the mouse hypomorphic *cobblestone* allele of another anterograde IFT gene, IFT88. In this mutant, cells mix freely between tel- and diencephalon and the border between the forebrain parts is dissolved (Willaredt et al. 2008). This feature is reminiscent of a failure to establish *zli*-mediated compartmentalization and consequently correct cellular segregation within the thalamus area. In *X. laevis,* depletion of *forkhead box J1 (foxj1)*, a TF regulating the biogenesis of motile cilia, induces striking forebrain defects (Hagenlocher et al. 2013). While normally in *Xenopus zli* cells cilia are longer than other primary cilia, in *foxj1* morphants the *zli* cilia are shortened and diencephalon size is massively reduced (Hagenlocher et al. 2013). *ftm/rpgrip1l* is a basal body protein involved in cilia structure and function. *Ftm* is a causal gene in severe human ciliopathies with brain abnormalities, including the Meckel–Gruber syndrome and Joubert syndrome of type B (Arts et al. 2007, Delous et al. 2007). Mouse mutant for ftm/*rpgrip1* locus exhibit a reduced number of cilia in the diencephalon. The remaining cilia are longer than in control embryos and display an abnormal shape. In the $ftm^{-/-}$ embryos the diencephalic territories are severely disorganized. Whereas in these embryos *shh* expression is mostly extinguished in the ventral forebrain, the *zli* territory is expanded. Despite Gli^A activity is drastically reduced in areas adjacent to the *zli* indicating that Shh read out is defective in $ftm^{-/-}$ embryos (Andreu-Cervera et al. 2019).

In conclusion, at this date diencephalic primordium, or *zli,* developmental defects have not been connected to thalamic fusion defects in human patients. Further studies are necessary to find out whether diencephalic mis-patterning may affect separation of the thalamic complex and contribute to the etiology of HPE. Failure to establish the anterior medial source of Shh, a lack of competence in the presumptive *zli* region, defective spreading and/or signaling of Shh could all contribute to alterations in formation of the p2/p3 border, the *zli,* and correct Shh, Notch and Fgf signaling in the thalamic field. All these events could participate to brain malformations.

CONCLUSION

The morphogenetic events underlying development of the vertebrate brain are sustained by a handful of extracellular signaling HH, Wnt, BMP, Fgf, Nodal, and Retinoid Acid (RA). The patterning and growth of the diencephalon is, in that sense, particularly meaningful as Shh together with Wnt, Fgf, and RA signaling act in a chronology- and topology- regulated manner and orchestrate the development and neurogenesis of the caudal forebrain territory (reviewed in (Martinez-Ferre and Martinez 2012, Epstein 2012, Hagemann and Scholpp 2012, Scholpp and Lumsden 2010)).

A large amount of work performed in the last twenty years provided important information on the sequential and coordinated events allowing emergence of the zli. More recently the *zli* developmental *cis-* and *trans-*determinants have been partially identified. Despite important information is still missing. The cues driving *zli* positioning, the identity of membrane proteins, probably Cadherins, allowing segregation of *zli* cells, and the direct or indirect role of the thalamus including the contribution of *otx, barhl, irx* and *pax6* to *zli* cellular properties, are still questioned. It will also be interesting to assess how the Shh signal is spread and read in the developing *zli,* and which signals generate the splitting of the midline in the diencephalic region, knowing that these processes are defective in HPE and ciliopathies. In the years to come, an important focus should be to better understand the gene-environment interactions involved in thalamic organogenesis and to develop animal models (chicken, zebrafish, mouse, *Xenopus*) mimicking human thalamic developmental disorders, a strategy that shall provide critical help into possible clinical interventions.

ACKNOWLEDGMENTS

We thank Kerstin Feistel for comments and discussion and Paul Johnson for his editing work on the manuscript. Sophie Portable of Sorbonne Université for illustrations. B Durand is funded by the Centre National de la Recherche Scientifique (CNRS). Her work was supported by the Fondation Pierre Gilles de Gennes (FPGG0039) and the Ligue Nationale contre le Cancer Comité Ile de France (RS19/75-52).

Conflicts of Interest

The authors declare no conflict of interest. The funding sponsors had neither a role in the design of the study, nor in the collection, analyses, or interpretation of data, nor in the writing of the manuscript, nor in the decision to publish the results.

REFERENCES

Acampora, D., M. P. Postiglione, V. Avantaggiato, M. Di Bonito, and A. Simeone. 2000. "The role of Otx and Otp genes in brain development." *Int J Dev Biol* 44 (6):669-77.

Alexandre, P., and M. Wassef. 2003. "The isthmic organizer links anteroposterior and dorsoventral patterning in the mid/hindbrain by generating roof plate structures." *Development* 130 (22):5331-8. doi: 10.1242/dev.00756.

Anderson, C., and C. D. Stern. 2016. "Organizers in Development." *Curr Top Dev Biol* 117:435-54. doi: 10.1016/bs.ctdb.2015.11.023.

Andreu-Cervera, A., I. Anselme, A. Karam, C. Laclef, M. Catala, and S. Schneider-Maunoury. 2019. "The Ciliopathy Gene Ftm/Rpgrip1l Controls Mouse Forebrain Patterning via Region-Specific Modulation of Hedgehog/Gli Signaling." *J Neurosci* 39 (13):2398-2415. doi: 10.1523/JNEUROSCI.2199-18.2019.

Arts, H. H., D. Doherty, S. E. van Beersum, M. A. Parisi, S. J. Letteboer, N. T. Gorden, T. A. Peters, T. Marker, K. Voesenek, A. Kartono, H. Ozyurek, F. M. Farin, H. Y. Kroes, U. Wolfrum, H. G. Brunner, F. P. Cremers, I. A. Glass, N. V. Knoers, and R. Roepman. 2007. "Mutations in the gene encoding the basal body protein RPGRIP1L, a nephrocystin-4 interactor, cause Joubert syndrome." *Nat Genet* 39 (7):882-8. doi: 10.1038/ng2069.

Baek, J. H., J. Hatakeyama, S. Sakamoto, T. Ohtsuka, and R. Kageyama. 2006. "Persistent and high levels of Hes1 expression regulate boundary formation in the developing central nervous system." *Development* 133 (13):2467-76. doi: 10.1242/dev.02403.

Bai, C. B., D. Stephen, and A. L. Joyner. 2004. "All mouse ventral spinal cord patterning by hedgehog is Gli dependent and involves an activator function of Gli3." *Dev Cell* 6 (1):103-15.

Bangs, F., and K. V. Anderson. 2017. "Primary Cilia and Mammalian Hedgehog Signaling." *Cold Spring Harb Perspect Biol* 9 (5). doi: 10.1101/cshperspect.a028175.

Barrios, N., E. Gonzalez-Perez, R. Hernandez, and S. Campuzano. 2015. "The Homeodomain Iroquois Proteins Control Cell Cycle Progression and Regulate the Size of Developmental Fields." *PLoS Genet* 11 (8):e1005463. doi: 10.1371/journal.pgen.1005463.

Beby, F., and T. Lamonerie. 2013. "The homeobox gene Otx2 in development and disease." *Exp Eye Res* 111:9-16. doi: 10.1016/j.exer.2013.03.007.

Beccari, L., R. Marco-Ferreres, and P. Bovolenta. 2013. "The logic of gene regulatory networks in early vertebrate forebrain patterning." *Mech Dev* 130 (2-3):95-111. doi: 10.1016/j.mod.2012.10.004.

Beddington, R. S., and E. J. Robertson. 1999. "Axis development and early asymmetry in mammals." *Cell* 96 (2):195-209.

Bergquist, H., and B. Kallen. 1954. "Notes on the early histogenesis and morphogenesis of the central nervous system in vertebrates." *J Comp Neurol* 100 (3):627-59.

Besse, L., M. Neti, I. Anselme, C. Gerhardt, U. Ruther, C. Laclef, and S. Schneider-Maunoury. 2011. "Primary cilia control telencephalic patterning and morphogenesis via Gli3 proteolytic processing." *Development* 138 (10):2079-88. doi: 10.1242/dev.059808.

Briscoe, J. 2009. "Making a grade: Sonic Hedgehog signalling and the control of neural cell fate." *EMBO J* 28 (5):457-65. doi: 10.1038/emboj.2009.12.

Buglino, J. A., and M. D. Resh. 2008. "Hhat is a palmitoylacyltransferase with specificity for N-palmitoylation of Sonic Hedgehog." *J Biol Chem* 283 (32):22076-88. doi: 10.1074/jbc.M803901200.

Bulfone, A., L. Puelles, M. H. Porteus, M. A. Frohman, G. R. Martin, and J. L. Rubenstein. 1993. "Spatially restricted expression of Dlx-1, Dlx-2 (Tes-1), Gbx-2, and Wnt-3 in the embryonic day 12.5 mouse forebrain defines potential transverse and longitudinal segmental boundaries." *J Neurosci* 13 (7):3155-72.

Bumcrot, D. A., R. Takada, and A. P. McMahon. 1995. "Proteolytic processing yields two secreted forms of sonic hedgehog." *Mol Cell Biol* 15 (4):2294-303.

Burke, R., D. Nellen, M. Bellotto, E. Hafen, K. A. Senti, B. J. Dickson, and K. Basler. 1999. "Dispatched, a novel sterol-sensing domain protein dedicated to the release of cholesterol-modified hedgehog from signaling cells." *Cell* 99 (7):803-15.

Caballero, I. M., M. N. Manuel, M. Molinek, I. Quintana-Urzainqui, D. Mi, T. Shimogori, and D. J. Price. 2014. "Cell-autonomous repression of Shh by transcription factor Pax6 regulates diencephalic patterning by controlling the central diencephalic organizer." *Cell Rep* 8 (5):1405-18. doi: 10.1016/j.celrep.2014.07.051.

Callejo, A., A. Bilioni, E. Mollica, N. Gorfinkiel, G. Andres, C. Ibanez, C. Torroja, L. Doglio, J. Sierra, and I. Guerrero. 2011. "Dispatched mediates Hedgehog basolateral release to form the long-range morphogenetic gradient in the Drosophila wing disk epithelium." *Proc Natl Acad Sci U S A* 108 (31):12591-8. doi: 10.1073/pnas.1106881108.

Cavodeassi, F., and C. Houart. 2012. "Brain regionalization: of signaling centers and boundaries." *Dev Neurobiol* 72 (3):218-33. doi: 10.1002/dneu.20938.

Cavodeassi, F., J. Modolell, and J. L. Gomez-Skarmeta. 2001. "The Iroquois family of genes: from body building to neural patterning." *Development* 128 (15):2847-55.

Chamberlain, C. E., J. Jeong, C. Guo, B. L. Allen, and A. P. McMahon. 2008. "Notochord-derived Shh concentrates in close association with the apically positioned basal body in neural target cells and forms a dynamic gradient during neural patterning." *Development* 135 (6):1097-106. doi: 10.1242/dev.013086.

Chamoun, Z., R. K. Mann, D. Nellen, D. P. von Kessler, M. Bellotto, P. A. Beachy, and K. Basler. 2001. "Skinny hedgehog, an acyltransferase required for palmitoylation and activity of the hedgehog signal." *Science* 293 (5537):2080-4. doi: 10.1126/science.1064437.

Chatterjee, M., Q. Guo, S. Weber, S. Scholpp, and J. Y. Li. 2014. "Pax6 regulates the formation of the habenular nuclei by controlling the temporospatial expression of Shh in the diencephalon in vertebrates." *BMC Biol* 12:13. doi: 10.1186/1741-7007-12-13.

Chatterjee, M., and J. Y. Li. 2012. "Patterning and compartment formation in the diencephalon." *Front Neurosci* 6:66. doi: 10.3389/fnins.2012.00066.

Chen, M. H., Y. J. Li, T. Kawakami, S. M. Xu, and P. T. Chuang. 2004. "Palmitoylation is required for the production of a soluble multimeric Hedgehog protein complex and long-range signaling in vertebrates." *Genes Dev* 18 (6):641-59. doi: 10.1101/gad.1185804.

Chen, X., H. Tukachinsky, C. H. Huang, C. Jao, Y. R. Chu, H. Y. Tang, B. Mueller, S. Schulman, T. A. Rapoport, and A. Salic. 2011. "Processing and turnover of the Hedgehog protein in the endoplasmic reticulum." *J Cell Biol* 192 (5):825-38. doi: 10.1083/jcb.201008090.

Cheng, Y. C., M. Amoyel, X. Qiu, Y. J. Jiang, Q. Xu, and D. G. Wilkinson. 2004. "Notch activation regulates the segregation and differentiation of rhombomere boundary cells in the zebrafish hindbrain." *Dev Cell* 6 (4):539-50.

Chenn, A., and C. A. Walsh. 2002. "Regulation of cerebral cortical size by control of cell cycle exit in neural precursors." *Science* 297 (5580):365-9.

Christ, A., A. Christa, E. Kur, O. Lioubinski, S. Bachmann, T. E. Willnow, and A. Hammes. 2012. "LRP2 is an auxiliary SHH receptor required to condition the forebrain ventral midline for inductive signals." *Dev Cell* 22 (2):268-78. doi: 10.1016/j.devcel.2011.11.023.

Chung, M. I., S. M. Peyrot, S. LeBoeuf, T. J. Park, K. L. McGary, E. M. Marcotte, and J. B. Wallingford. 2012. "RFX2 is broadly required for ciliogenesis during vertebrate development." *Dev Biol* 363 (1):155-65. doi: 10.1016/j.ydbio.2011.12.029.

Cinnamon, E., and Z. Paroush. 2008. "Context-dependent regulation of Groucho/TLE-mediated repression." *Curr Opin Genet Dev* 18 (5):435-40. doi: 10.1016/j.gde.2008.07.010.

Coggeshall, R. E. 1964. "A Study of Diencephalic Development in the Albino Rat." *J Comp Neurol* 122:241-69.

Colombo, A., G. Reig, M. Mione, and M. L. Concha. 2006. "Zebrafish BarH-like genes define discrete neural domains in the early embryo." *Gene Expr Patterns* 6 (4):347-52.

Couly, G. F., and N. M. Le Douarin. 1987. "Mapping of the early neural primordium in quail-chick chimeras. II. The prosencephalic neural plate and neural folds: implications for the genesis of cephalic human congenital abnormalities." *Dev Biol* 120 (1):198-214.

Creanga, A., T. D. Glenn, R. K. Mann, A. M. Saunders, W. S. Talbot, and P. A. Beachy. 2012. "Scube/You activity mediates release of dually lipid-modified Hedgehog signal in soluble form." *Genes Dev* 26 (12):1312-25. doi: 10.1101/gad.191866.112.

Crossley, P. H., and G. R. Martin. 1995. "The mouse Fgf8 gene encodes a family of polypeptides and is expressed in regions that direct outgrowth and patterning in the developing embryo." *Development* 121 (2):439-51.

Crossley, P. H., S. Martinez, and G. R. Martin. 1996. "Midbrain development induced by FGF8 in the chick embryo." *Nature* 380 (6569):66-8.

Cruz, C., V. Ribes, E. Kutejova, J. Cayuso, V. Lawson, D. Norris, J. Stevens, M. Davey, K. Blight, F. Bangs, A. Mynett, E. Hirst, R. Chung, N. Balaskas, S. L. Brody, E. Marti, and J. Briscoe. 2010. "Foxj1 regulates floor plate cilia architecture and modifies the response of cells to sonic hedgehog signalling." *Development* 137 (24):4271-82. doi: 10.1242/dev.051714.

Dahmann, C., A. C. Oates, and M. Brand. 2011. "Boundary formation and maintenance in tissue development." *Nat Rev Genet* 12 (1):43-55. doi: 10.1038/nrg2902.

de la Calle-Mustienes, E., J. Modolell, and J. L. Gomez-Skarmeta. 2002. "The Xiro-repressed gene CoREST is expressed in Xenopus neural territories." *Mech Dev* 110 (1-2):209-11.

De Robertis, E. M., J. Larrain, M. Oelgeschlager, and O. Wessely. 2000. "The establishment of Spemann's organizer and patterning of the vertebrate embryo." *Nat Rev Genet* 1 (3):171-81. doi: 10.1038/35042039.

Delous, M., L. Baala, R. Salomon, C. Laclef, J. Vierkotten, K. Tory, C. Golzio, T. Lacoste, L. Besse, C. Ozilou, I. Moutkine, N. E. Hellman, I. Anselme, F. Silbermann, C. Vesque, C. Gerhardt, E. Rattenberry, M. T. Wolf, M. C. Gubler, J. Martinovic, F. Encha-Razavi, N. Boddaert, M. Gonzales, M. A. Macher, H. Nivet, G. Champion, J. P. Bertheleme, P. Niaudet, F. McDonald, F. Hildebrandt, C. A. Johnson, M. Vekemans, C. Antignac, U. Ruther, S. Schneider-Maunoury, T. Attie-Bitach, and S. Saunier. 2007. "The ciliary gene RPGRIP1L is mutated in cerebello-oculo-renal syndrome (Joubert syndrome type B) and Meckel syndrome." *Nat Genet* 39 (7):875-81. doi: 10.1038/ng2039.

Ding, Q., R. Balasubramanian, D. Zheng, G. Liang, and L. Gan. 2016. "Barhl2 Determines the Early Patterning of the Diencephalon by Regulating Shh." *Mol Neurobiol*. doi: 10.1007/s12035-016-0001-5.

Dreesen, O., and A. H. Brivanlou. 2007. "Signaling pathways in cancer and embryonic stem cells." *Stem Cell Rev* 3 (1):7-17.

Dubourg, C., W. Carre, H. Hamdi-Roze, C. Mouden, J. Roume, B. Abdelmajid, D. Amram, C. Baumann, N. Chassaing, C. Coubes, L. Faivre-Olivier, E. Ginglinger, M. Gonzales, A. Levy-Mozziconacci, S. A. Lynch, S. Naudion, L. Pasquier, A. Poidvin, F. Prieur, P. Sarda, A. Toutain, V. Dupe, L. Akloul, S. Odent, M. de Tayrac, and V. David. 2016. "Mutational Spectrum in Holoprosencephaly Shows That Fgf is a New Major Signaling Pathway." *Hum Mutat*. doi: 10.1002/humu.23038.

Dupe, V., L. Rochard, S. Mercier, Y. Le Petillon, I. Gicquel, C. Bendavid, G. Bourrouillou, U. Kini, C. Thauvin-Robinet, T. P. Bohan, S. Odent, C. Dubourg, and V. David. 2011. "NOTCH, a new signaling pathway implicated in holoprosencephaly." *Hum Mol Genet* 20 (6):1122-31. doi: 10.1093/hmg/ddq556.

Durand, B. C. 2016. "Stem cell-like Xenopus Embryonic Explants to Study Early Neural Developmental Features In Vitro and In Vivo." *J Vis Exp* (108):e53474. doi: 10.3791/53474.

Eagleson, G. W., and W. A. Harris. 1990. "Mapping of the presumptive brain regions in the neural plate of Xenopus laevis." *J Neurobiol* 21 (3):427-40.

Echevarri, D. a, C. Vieira, L. Gimeno, and S. Martinez. 2003. "Neuroepithelial secondary organizers and cell fate specification in the developing brain." *Brain Res Brain Res Rev* 43 (2):179-191.

Epstein, D. J. 2012. "Regulation of thalamic development by sonic hedgehog." *Front Neurosci* 6:57.

Ericson, J., S. Morton, A. Kawakami, H. Roelink, and T. M. Jessell. 1996. "Two critical periods of Sonic Hedgehog signaling required for the specification of motor neuron identity." *Cell* 87 (4):661-73.

Eugster, C., D. Panakova, A. Mahmoud, and S. Eaton. 2007. "Lipoprotein-heparan sulfate interactions in the Hh pathway." *Dev Cell* 13 (1):57-71. doi: 10.1016/j.devcel.2007.04.019.

Fernandes, M., and J. M. Hebert. 2008. "The ups and downs of holoprosencephaly: dorsal versus ventral patterning forces." *Clin Genet* 73 (5):413-23. doi: 10.1111/j.1399-0004.2008.00994.x.

Ferran, J. L., L. Puelles, and J. L. Rubenstein. 2015. "Molecular codes defining rostrocaudal domains in the embryonic mouse hypothalamus." *Front Neuroanat* 9:46. doi: 10.3389/fnana.2015.00046.

Figdor, M. C., and C. D. Stern. 1993. "Segmental organization of embryonic diencephalon." *Nature* 363 (6430):630-4. doi: 10.1038/363630a0.

Fuccillo, M., A. L. Joyner, and G. Fishell. 2006. "Morphogen to mitogen: the multiple roles of hedgehog signalling in vertebrate neural development." *Nat Rev Neurosci* 7 (10):772-83. doi: 10.1038/nrn1990.

Garcia-Lopez, R., C. Vieira, D. Echevarria, and S. Martinez. 2004. "Fate map of the diencephalon and the zona limitans at the 10-somites stage in chick embryos." *Dev Biol* 268 (2):514-30.

Gomez-Skarmeta, J. L., S. Campuzano, and J. Modolell. 2003. "Half a century of neural prepatterning: the story of a few bristles and many genes." *Nat Rev Neurosci* 4 (7):587-98.

Gomez-Skarmeta, J. L., and J. Modolell. 2002. "Iroquois genes: genomic organization and function in vertebrate neural development." *Curr Opin Genet Dev* 12 (4):403-8. doi: S0959437X02003179 [pii].

Gongal, P. A., C. R. French, and A. J. Waskiewicz. 2011. "Aberrant forebrain signaling during early development underlies the generation of holoprosencephaly and coloboma." *Biochim Biophys Acta* 1812 (3):390-401. doi: 10.1016/j.bbadis.2010.09.005.

Gorivodsky, M., M. Mukhopadhyay, M. Wilsch-Braeuninger, M. Phillips, A. Teufel, C. Kim, N. Malik, W. Huttner, and H. Westphal. 2009. "Intraflagellar transport protein 172 is essential for primary cilia formation and plays a vital role in patterning the mammalian brain." *Dev Biol* 325 (1):24-32. doi: 10.1016/j.ydbio.2008.09.019.

Grigoryan, T., P. Wend, A. Klaus, and W. Birchmeier. 2008. "Deciphering the function of canonical Wnt signals in development and disease: conditional loss- and gain-of-function mutations of beta-catenin in mice." *Genes Dev* 22 (17):2308-41.

Guerrini, R., and W. B. Dobyns. 2014. "Malformations of cortical development: clinical features and genetic causes." *Lancet Neurol* 13 (7):710-26. doi: 10.1016/S1474-4422(14)70040-7.

Hagemann, A. I., and S. Scholpp. 2012. "The Tale of the Three Brothers - Shh, Wnt, and Fgf during Development of the Thalamus." *Front Neurosci* 6:76.

Hagenlocher, C., P. Walentek, M. Ller C, T. Thumberger, and K. Feistel. 2013. "Ciliogenesis and cerebrospinal fluid flow in the developing Xenopus brain are regulated by foxj1." *Cilia* 2 (1):12. doi: 10.1186/2046-2530-2-12.

Hahn, J. S., and P. D. Barnes. 2010. "Neuroimaging advances in holoprosencephaly: Refining the spectrum of the midline malformation." *Am J Med Genet C Semin Med Genet* 154C (1):120-32. doi: 10.1002/ajmg.c.30238.

Hardy, R. Y., and M. D. Resh. 2012. "Identification of N-terminal residues of Sonic Hedgehog important for palmitoylation by Hedgehog acyltransferase." *J Biol Chem* 287 (51):42881-9. doi: 10.1074/jbc.M112.426833.

Harland, R. 2000. "Neural induction." *Curr Opin Genet Dev* 10 (4):357-62. doi: S0959-437X(00)00096-4 [pii].

Hashimoto-Torii, K., J. Motoyama, C. C. Hui, A. Kuroiwa, M. Nakafuku, and K. Shimamura. 2003. "Differential activities of Sonic hedgehog mediated by Gli transcription factors define distinct neuronal subtypes in the dorsal thalamus." *Mech Dev* 120 (10):1097-111.

Heasman, J. 2006. "Patterning the early Xenopus embryo." *Development* 133 (7):1205-17. doi: 10.1242/dev.02304.

Hebert, J. M., and G. Fishell. 2008. "The genetics of early telencephalon patterning: some assembly required." *Nat Rev Neurosci* 9 (9):678-85. doi: 10.1038/nrn2463.

Hilgendorf, K. I., C. T. Johnson, and P. K. Jackson. 2016. "The primary cilium as a cellular receiver: organizing ciliary GPCR signaling." *Curr Opin Cell Biol* 39:84-92. doi: 10.1016/j.ceb.2016.02.008.

Hirata, T., M. Nakazawa, O. Muraoka, R. Nakayama, Y. Suda, and M. Hibi. 2006. "Zinc-finger genes Fez and Fez-like function in the establishment of diencephalon subdivisions." *Development* 133 (20):3993-4004.

Hoch, R. V., J. L. Rubenstein, and S. Pleasure. 2009. "Genes and signaling events that establish regional patterning of the mammalian forebrain." *Semin Cell Dev Biol* 20 (4):378-86.

Holzschuh, J., G. Hauptmann, and W. Driever. 2003. "Genetic analysis of the roles of Hh, FGF8, and nodal signaling during catecholaminergic system development in the zebrafish brain." *J Neurosci* 23 (13):5507-19.

Houart, C., M. Westerfield, and S. W. Wilson. 1998. "A small population of anterior cells patterns the forebrain during zebrafish gastrulation." *Nature* 391 (6669):788-92.

Hu, W. F., M. H. Chahrour, and C. A. Walsh. 2014. "The diverse genetic landscape of neurodevelopmental disorders." *Annu Rev Genomics Hum Genet* 15:195-213. doi: 10.1146/annurev-genom-090413-025600.

Huang, A. S., J. A. Mitchell, S. N. Haber, N. Alia-Klein, and R. Z. Goldstein. 2018. "The thalamus in drug addiction: from rodents to humans." *Philos Trans R Soc Lond B Biol Sci* 373 (1742). doi: 10.1098/rstb.2017.0028.

Huangfu, D., A. Liu, A. S. Rakeman, N. S. Murcia, L. Niswander, and K. V. Anderson. 2003. "Hedgehog signalling in the mouse requires intraflagellar transport proteins." *Nature* 426 (6962):83-7. doi: 10.1038/nature02061.

Jacob, J., J. Briscoe, J. Britto, D. Tannahill, and R. Keynes. 2003. "Gli proteins and the control of spinal-cord patterning." *EMBO Rep* 4 (8):761-5. CB2 3DY, UK.

Jennings, B. H., and D. Ish-Horowicz. 2008. "The Groucho/TLE/Grg family of transcriptional co-repressors." *Genome Biol* 9 (1):205. doi: 10.1186/gb-2008-9-1-205.

Jeong, J. Y., Z. Einhorn, P. Mathur, L. Chen, S. Lee, K. Kawakami, and S. Guo. 2007. "Patterning the zebrafish diencephalon by the conserved zinc-finger protein Fezl." *Development* 134 (1):127-36.

Jeong, Y., D. K. Dolson, R. R. Waclaw, M. P. Matise, L. Sussel, K. Campbell, K. H. Kaestner, and D. J. Epstein. 2011. "Spatial and temporal requirements for sonic hedgehog in the regulation of thalamic interneuron identity." *Development* 138 (3):531-41. doi: 10.1242/dev.058917.

Jessell, T. M. 2000. "Neuronal specification in the spinal cord: inductive signals and transcriptional codes." *Nat Rev Genet* 1 (1):20-9.

Jessell, T. M., and J. R. Sanes. 2000. "Development. The decade of the developing brain." *Curr Opin Neurobiol* 10 (5):599-611.

Jimenez-Guri, E., and C. Pujades. 2011. "An ancient mechanism of hindbrain patterning has been conserved in vertebrate evolution." *Evol Dev* 13 (1):38-46. doi: 10.1111/j.1525-142X.2010.00454.x.

Juraver-Geslin, H. A., J. J. Ausseil, M. Wassef, and B. C. Durand. 2011. "Barhl2 limits growth of the diencephalic primordium through Caspase3 inhibition of beta-catenin activation." *Proc Natl Acad Sci U S A* 108 (6):2288-93. doi: 10.1073/pnas.1014017108.

Juraver-Geslin, H. A., and B. C. Durand. 2015. "Early development of the neural plate: new roles for apoptosis and for one of its main effectors caspase-3." *Genesis* 53 (2):203-24. doi: 10.1002/dvg.22844.

Juraver-Geslin, H. A., J. L. Gomez-Skarmeta, and B. C. Durand. 2014. "The conserved barH-like homeobox-2 gene barhl2 acts downstream of orthodentricle-2 and together with iroquois-3 in establishment of the caudal forebrain signaling center induced by Sonic Hedgehog." *Dev Biol* 396 (1):107-20. doi: 10.1016/j.ydbio.2014.09.027.

Keyser, A. 1972. "The development of the diencephalon of the Chinese hamster. An investigation of the validity of the criteria of subdivision of the brain." *Acta Anat Suppl (Basel)* 59:1-178.

Kiecker, C., and A. Lumsden. 2004. "Hedgehog signaling from the ZLI regulates diencephalic regional identity." *Nat Neurosci* 7 (11):1242-9.

Kiecker, C., and A. Lumsden. 2005. "Compartments and their boundaries in vertebrate brain development." *Nat Rev Neurosci* 6 (7):553-64.

Kiecker, C., and A. Lumsden. 2012. "The role of organizers in patterning the nervous system." *Annu Rev Neurosci* 35:347-67. doi: 10.1146/annurev-neuro-062111-150543.

Kobayashi, D., M. Kobayashi, K. Matsumoto, T. Ogura, M. Nakafuku, and K. Shimamura. 2002. "Early subdivisions in the neural plate define distinct competence for inductive signals." *Development* 129 (1):83-93.

Kohtz, J. D., H. Y. Lee, N. Gaiano, J. Segal, E. Ng, T. Larson, D. P. Baker, E. A. Garber, K. P. Williams, and G. Fishell. 2001. "N-terminal fatty-acylation of sonic hedgehog enhances the induction of rodent ventral forebrain neurons." *Development* 128 (12):2351-63.

Kramer-Zucker, A. G., F. Olale, C. J. Haycraft, B. K. Yoder, A. F. Schier, and I. A. Drummond. 2005. "Cilia-driven fluid flow in the zebrafish pronephros, brain and Kupffer's vesicle is required for normal organogenesis." *Development* 132 (8):1907-21. doi: 10.1242/dev.01772.

Kraus, Y., A. Aman, U. Technau, and G. Genikhovich. 2016. "Pre-bilaterian origin of the blastoporal axial organizer." *Nat Commun* 7:11694. doi: 10.1038/ncomms11694.

Laclef, C., I. Anselme, L. Besse, M. Catala, A. Palmyre, D. Baas, M. Paschaki, M. Pedraza, C. Metin, B. Durand, and S. Schneider-Maunoury. 2015. "The role of primary cilia in corpus callosum formation is mediated by production of the Gli3 repressor." *Hum Mol Genet* 24 (17):4997-5014. doi: 10.1093/hmg/ddv221.

Larsen, C. W., L. M. Zeltser, and A. Lumsden. 2001. "Boundary formation and compartition in the avian diencephalon." *J Neurosci* 21 (13):4699-711.

Lecaudey, V., I. Anselme, F. Rosa, and S. Schneider-Maunoury. 2004. "The zebrafish Iroquois gene iro7 positions the r4/r5 boundary and controls neurogenesis in the rostral hindbrain." *Development* 131 (13):3121-31. doi: 10.1242/dev.01190. dev.01190 [pii].

Li, Y., H. Zhang, Y. Litingtung, and C. Chiang. 2006. "Cholesterol modification restricts the spread of Shh gradient in the limb bud." *Proc Natl Acad Sci U S A* 103 (17):6548-53. doi: 10.1073/pnas.0600124103.

Liegeois, S., A. Benedetto, J. M. Garnier, Y. Schwab, and M. Labouesse. 2006. "The V0-ATPase mediates apical secretion of exosomes containing Hedgehog-related proteins in Caenorhabditis elegans." *J Cell Biol* 173 (6):949-61. doi: 10.1083/jcb.200511072.

Llinas, R. R. 2003. "[Thalamo-cortical dysrhythmia syndrome: neuropsychiatric features]." *An R Acad Nac Med (Madr)* 120 (2):267-90; discussion 290-5.

Lowe, C. J., M. Wu, A. Salic, L. Evans, E. Lander, N. Stange-Thomann, C. E. Gruber, J. Gerhart, and M. Kirschner. 2003. "Anteroposterior patterning in hemichordates and the origins of the chordate nervous system." *Cell* 113 (7):853-65.

Lumsden, A., and R. Keynes. 1989. "Segmental patterns of neuronal development in the chick hindbrain." *Nature* 337 (6206):424-8. doi: 10.1038/337424a0.

Marcorelles, P., and A. Laquerriere. 2010. "Neuropathology of holoprosencephaly." *Am J Med Genet C Semin Med Genet* 154C (1):109-19. doi: 10.1002/ajmg.c.30249.

Martinez, S., and L. Puelles. 2000. "Neurogenetic compartments of the mouse diencephalon and some characteristic gene expression patterns." *Results Probl Cell Differ* 30:91-106.

Martinez-Ferre, A., M. Navarro-Garberi, C. Bueno, and S. Martinez. 2013. "Wnt signal specifies the intrathalamic limit and its organizer properties by regulating Shh induction in the alar plate." *J Neurosci* 33 (9):3967-80. doi: 10.1523/JNEUROSCI.0726-12.2013.

Martinez-Ferre, Almudena, and Salvador Martinez. 2012. "Molecular regionalization of the diencephalon." *Frontiers In Neuroscience* 6:73-73. doi: 10.3389/fnins.2012.00073.

Mathieu, J., A. Barth, F. M. Rosa, S. W. Wilson, and N. Peyrieras. 2002. "Distinct and cooperative roles for Nodal and Hedgehog signals during hypothalamic development." *Development* 129 (13):3055-65.

Mattes, B., S. Weber, J. Peres, Q. Chen, G. Davidson, C. Houart, and S. Scholpp. 2012. "Wnt3 and Wnt3a are required for induction of the mid-diencephalic organizer in the caudal forebrain." *Neural Dev* 7:12.

Megason, S. G., and A. P. McMahon. 2002. "A mitogen gradient of dorsal midline Wnts organizes growth in the CNS." *Development* 129 (9):2087-98.

Mehlen, P., F. Mille, and C. Thibert. 2005. "Morphogens and cell survival during development." *J Neurobiol* 64 (4):357-66. doi: 10.1002/neu.20167.

Mercier, S., V. David, L. Ratie, I. Gicquel, S. Odent, and V. Dupe. 2013. "NODAL and SHH dose-dependent double inhibition promotes an HPE-like phenotype in chick embryos." *Dis Model Mech* 6 (2):537-43. doi: 10.1242/dmm.010132.

Mo, Z., S. Li, X. Yang, and M. Xiang. 2004. "Role of the Barhl2 homeobox gene in the specification of glycinergic amacrine cells." *Development* 131 (7):1607-18.

Muenke, M., and P. A. Beachy. 2000. "Genetics of ventral forebrain development and holoprosencephaly." *Curr Opin Genet Dev* 10 (3):262-9. doi: S0959-437X(00)00084-8 [pii].

Mukhopadhyay, S., and R. Rohatgi. 2014. "G-protein-coupled receptors, Hedgehog signaling and primary cilia." *Semin Cell Dev Biol* 33:63-72. doi: 10.1016/j.semcdb.2014.05.002.

Munke, M. 1989. "Clinical, cytogenetic, and molecular approaches to the genetic heterogeneity of holoprosencephaly." *Am J Med Genet* 34 (2):237-45. doi: 10.1002/ajmg.1320340222.

Nanni, L., J. E. Ming, M. Bocian, K. Steinhaus, D. W. Bianchi, C. Die-Smulders, A. Giannotti, K. Imaizumi, K. L. Jones, M. D. Campo, R. A. Martin, P. Meinecke, M. E. Pierpont, N. H. Robin, I. D. Young, E. Roessler, and M. Muenke. 1999. "The mutational spectrum of the sonic hedgehog gene in holoprosencephaly: SHH mutations cause a significant proportion of autosomal dominant holoprosencephaly." *Hum Mol Genet* 8 (13):2479-88.

Niehrs, C. 2004. "Regionally specific induction by the Spemann-Mangold organizer." *Nat Rev Genet* 5 (6):425-34. doi: 10.1038/nrg1347 [pii].

Noelanders, R., and K. Vleminckx. 2016. "How Wnt Signaling Builds the Brain: Bridging Development and Disease." *Neuroscientist*. doi: 10.1177/1073858416667270.

Offner, N., N. Duval, M. Jamrich, and B. Durand. 2005. "The pro-apoptotic activity of a vertebrate Bar-like homeobox gene plays a key role in patterning the Xenopus neural plate by limiting the number of chordin- and shh-expressing cells." *Development* 132 (8):1807-18.

Panakova, D., H. Sprong, E. Marois, C. Thiele, and S. Eaton. 2005. "Lipoprotein particles are required for Hedgehog and Wingless signalling." *Nature* 435 (7038):58-65. doi: 10.1038/nature03504.

Pani, A. M., E. E. Mullarkey, J. Aronowicz, S. Assimacopoulos, E. A. Grove, and C. J. Lowe. 2012. "Ancient deuterostome origins of vertebrate brain signalling centres." *Nature* 483 (7389):289-94. doi: 10.1038/nature10838.

Pannese, M., C. Polo, M. Andreazzoli, R. Vignali, B. Kablar, G. Barsacchi, and E. Boncinelli. 1995. "The Xenopus homologue of Otx2 is a maternal homeobox gene that demarcates and specifies anterior body regions." *Development* 121 (3):707-20.

Park, S. M., H. J. Jang, and J. H. Lee. 2019. "Roles of Primary Cilia in the Developing Brain." *Front Cell Neurosci* 13:218. doi: 10.3389/fncel.2019.00218.

Patterson, K. D., O. Cleaver, W. V. Gerber, F. G. White, and P. A. Krieg. 2000. "Distinct expression patterns for two Xenopus Bar homeobox genes." *Dev Genes Evol* 210 (3):140-4.

Pepinsky, R. B., C. Zeng, D. Wen, P. Rayhorn, D. P. Baker, K. P. Williams, S. A. Bixler, C. M. Ambrose, E. A. Garber, K. Miatkowski, F. R. Taylor, E. A. Wang, and A. Galdes. 1998. "Identification of a palmitic acid-modified form of human Sonic hedgehog." *J Biol Chem* 273 (22):14037-45.

Peters, T., R. Dildrop, K. Ausmeier, and U. Ruther. 2000. "Organization of mouse Iroquois homeobox genes in two clusters suggests a conserved regulation and function in vertebrate development." *Genome Res* 10 (10):1453-62.

Petryk, A., D. Graf, and R. Marcucio. 2015. "Holoprosencephaly: signaling interactions between the brain and the face, the environment and the genes, and the phenotypic variability in animal models and humans." *Wiley Interdiscip Rev Dev Biol* 4 (1):17-32. doi: 10.1002/wdev.161.

Peukert, D., S. Weber, A. Lumsden, and S. Scholpp. 2011. "Lhx2 and Lhx9 determine neuronal differentiation and compartition in the caudal forebrain by regulating Wnt signaling." *PLoS Biol* 9 (12):e1001218.

Placzek, M., and J. Briscoe. 2005. "The floor plate: multiple cells, multiple signals." *Nat Rev Neurosci* 6 (3):230-40.

Placzek, M., M. Tessier-Lavigne, T. Yamada, T. Jessell, and J. Dodd. 1990. "Mesodermal control of neural cell identity: floor plate induction by the notochord." *Science* 250 (4983):985-8.

Porter, J. A., S. C. Ekker, W. J. Park, D. P. von Kessler, K. E. Young, C. H. Chen, Y. Ma, A. S. Woods, R. J. Cotter, E. V. Koonin, and P. A. Beachy. 1996. "Hedgehog patterning activity: role of a lipophilic modification mediated by the carboxy-terminal autoprocessing domain." *Cell* 86 (1):21-34.

Porter, J. A., D. P. von Kessler, S. C. Ekker, K. E. Young, J. J. Lee, K. Moses, and P. A. Beachy. 1995. "The product of hedgehog autoproteolytic cleavage active in local and long-range signalling." *Nature* 374 (6520):363-6. doi: 10.1038/374363a0.

Porter, J. A., K. E. Young, and P. A. Beachy. 1996. "Cholesterol modification of hedgehog signaling proteins in animal development." *Science* 274 (5285):255-9.

Puelles, E. 2007. "Genetic control of basal midbrain development." *J Neurosci Res* 85 (16):3530-4.

Puelles, L. 2018. "Developmental studies of avian brain organization." *Int J Dev Biol* 62 (1-2-3):207-224. doi: 10.1387/ijdb.170279LP.

Puelles, L., M. Harrison, G. Paxinos, and C. Watson. 2013. "A developmental ontology for the mammalian brain based on the prosomeric model." *Trends Neurosci* 36 (10):570-8. doi: 10.1016/j.tins.2013.06.004.

Puelles, Luis, and John L. R. Rubenstein. 2003. "Forebrain gene expression domains and the evolving prosomeric model." *Trends in Neurosciences* 26:469-476. doi: 10.1016/S0166-2236(03)00234-0.

Raff, M. 1998. "Cell suicide for beginners." *Nature* 396 (6707):119-22.

Ramakrishnan, A. B., A. Sinha, V. B. Fan, and K. M. Cadigan. 2018. "The Wnt Transcriptional Switch: TLE Removal or Inactivation?" *Bioessays* 40 (2). doi: 10.1002/bies.201700162.

Ramsbottom, S. A., and M. E. Pownall. 2016. "Regulation of Hedgehog Signalling Inside and Outside the Cell." *J Dev Biol* 4 (3):23. doi: 10.3390/jdb4030023.

Redies, C. 2000. "Cadherins in the central nervous system." *Prog Neurobiol* 61 (6):611-48.

Reig, G., M. E. Cabrejos, and M. L. Concha. 2007. "Functions of BarH transcription factors during embryonic development." *Dev Biol* 302 (2):367-75.
Robertshaw, E., K. Matsumoto, A. Lumsden, and C. Kiecker. 2013. "Irx3 and Pax6 establish differential competence for Shh-mediated induction of GABAergic and glutamatergic neurons of the thalamus." *Proc Natl Acad Sci U S A* 110 (41):E3919-E3926. doi: 10.1073/pnas.1304311110.
Rodriguez-Seguel, E., P. Alarcon, and J. L. Gomez-Skarmeta. 2009. "The Xenopus Irx genes are essential for neural patterning and define the border between prethalamus and thalamus through mutual antagonism with the anterior repressors Fezf and Arx." *Dev Biol* 329 (2):258-68.
Rubenstein, J. L., S. Martinez, K. Shimamura, and L. Puelles. 1994. "The embryonic vertebrate forebrain: the prosomeric model." *Science* 266 (5185):578-80.
Rubenstein, J. L., K. Shimamura, S. Martinez, and L. Puelles. 1998. "Regionalization of the prosencephalic neural plate." *Annu Rev Neurosci* 21:445-77.
Ruiz i Altaba, A., V. Nguyen, and V. Palma. 2003. "The emergent design of the neural tube: prepattern, SHH morphogen and GLI code." *Curr Opin Genet Dev* 13 (5):513-21.
Sampath, K., A. L. Rubinstein, A. M. Cheng, J. O. Liang, K. Fekany, L. Solnica-Krezel, V. Korzh, M. E. Halpern, and C. V. Wright. 1998. "Induction of the zebrafish ventral brain and floorplate requires cyclops/nodal signalling." *Nature* 395 (6698):185-9. doi: 10.1038/26020.
Scheibel, A. B. 1997. "The thalamus and neuropsychiatric illness." *J Neuropsychiatry Clin Neurosci* 9 (3):342-53. doi: 10.1176/jnp.9.3.342.
Scholpp, S., I. Foucher, N. Staudt, D. Peukert, A. Lumsden, and C. Houart. 2007. "Otx1l, Otx2 and Irx1b establish and position the ZLI in the diencephalon." *Development* 134 (17):3167-76.
Scholpp, S., O. Wolf, M. Brand, and A. Lumsden. 2006. "Hedgehog signalling from the zona limitans intrathalamica orchestrates patterning of the zebrafish diencephalon." *Development* 133 (5):855-64.

Scholpp, Steffen, and Andrew Lumsden. 2010. "Review: Building a bridal chamber: development of the thalamus." *Trends in Neurosciences* 33:373-380. doi: 10.1016/j.tins.2010.05.003.

Schuhmacher, L. N., S. Albadri, M. Ramialison, and L. Poggi. 2011. "Evolutionary relationships and diversification of barhl genes within retinal cell lineages." *BMC Evol Biol* 11:340. doi: 10.1186/1471-2148-11-340.

Sena, E., N. Rocques, C. Borday, H. S. Muhamad Amin, K. Parain, D. Sitbon, A. Chesneau, and B. C. Durand. 2019. "Barhl2 maintains T cell factors as repressors and thereby switches off the Wnt/beta-Catenin response driving Spemann organizer formation." *Development* 146 (10). doi: 10.1242/dev.173112.

Shelton, L., L. Becerra, and D. Borsook. 2012. "Unmasking the mysteries of the habenula in pain and analgesia." *Prog Neurobiol* 96 (2):208-19. doi: 10.1016/j.pneurobio.2012.01.004.

Shepherd, G. M. 2013. "Corticostriatal connectivity and its role in disease." *Nat Rev Neurosci* 14 (4):278-91. doi: 10.1038/nrn3469.

Shimamura, K., D. J. Hartigan, S. Martinez, L. Puelles, and J. L. Rubenstein. 1995. "Longitudinal organization of the anterior neural plate and neural tube." *Development* 121 (12):3923-33.

Sjoqvist, M., and E. R. Andersson. 2019. "Do as I say, Not(ch) as I do: Lateral control of cell fate." *Dev Biol* 447 (1):58-70. doi: 10.1016/j.ydbio.2017.09.032.

Staudt, N., and C. Houart. 2007. "The prethalamus is established during gastrulation and influences diencephalic regionalization." *PLoS Biol* 5 (4):e69.

Stern, C. D. 2001. "Initial patterning of the central nervous system: how many organizers?" *Nat Rev Neurosci* 2 (2):92-8. doi: 10.1038/35053563.

Stern, C. D. 2002. "Induction and initial patterning of the nervous system - the chick embryo enters the scene." *Curr Opin Genet Dev* 12 (4):447-51.

Stern, C. D. 2005. "Neural induction: old problem, new findings, yet more questions." *Development* 132 (9):2007-21. doi: 10.1242/dev.01794.

Stern, C. D. 2006. "Neural induction: 10 years on since the 'default model'." *Curr Opin Cell Biol* 18 (6):692-7. doi: S0955-0674(06)00147-5 [pii] 10.1016/j.ceb.2006.09.002.

Szabo, N. E., T. Zhao, X. Zhou, and G. Alvarez-Bolado. 2009. "The role of Sonic hedgehog of neural origin in thalamic differentiation in the mouse." *J Neurosci* 29 (8):2453-66. doi: 29/8/2453 [pii] 10.1523/JNEUROSCI.4524-08.2009.

Thompson, C. L., L. Ng, V. Menon, S. Martinez, C. K. Lee, K. Glattfelder, S. M. Sunkin, A. Henry, C. Lau, C. Dang, R. Garcia-Lopez, A. Martinez-Ferre, A. Pombero, J. L. Rubenstein, W. B. Wakeman, J. Hohmann, N. Dee, A. J. Sodt, R. Young, K. Smith, T. N. Nguyen, J. Kidney, L. Kuan, A. Jeromin, A. Kaykas, J. Miller, D. Page, G. Orta, A. Bernard, Z. Riley, S. Smith, P. Wohnoutka, M. J. Hawrylycz, L. Puelles, and A. R. Jones. 2014. "A high-resolution spatiotemporal atlas of gene expression of the developing mouse brain." *Neuron* 83 (2):309-23. doi: 10.1016/j.neuron.2014.05.033.

Tossell, K., C. Kiecker, A. Wizenmann, E. Lang, and C. Irving. 2011. "Notch signalling stabilises boundary formation at the midbrain-hindbrain organiser." *Development* 138 (17):3745-57. doi: 10.1242/dev.070318.

Tukachinsky, H., R. P. Kuzmickas, C. Y. Jao, J. Liu, and A. Salic. 2012. "Dispatched and scube mediate the efficient secretion of the cholesterol-modified hedgehog ligand." *Cell Rep* 2 (2):308-20. doi: 10.1016/j.celrep.2012.07.010.

Vaage, S. 1969. "The segmentation of the primitive neural tube in chick embryos (Gallus domesticus). A morphological, histochemical and autoradiographical investigation." *Ergeb Anat Entwicklungsgesch* 41 (3):3-87.

van Amerongen, R., and R. Nusse. 2009. "Towards an integrated view of Wnt signaling in development." *Development* 136 (19):3205-14. doi: 10.1242/dev.033910.

Vieira, C., A. L. Garda, K. Shimamura, and S. Martinez. 2005. "Thalamic development induced by Shh in the chick embryo." *Dev Biol* 284 (2):351-63.

Vieira, C., and S. Martinez. 2006. "Sonic hedgehog from the basal plate and the zona limitans intrathalamica exhibits differential activity on diencephalic molecular regionalization and nuclear structure." *Neuroscience* 143 (1):129-40.

Villa-Cuesta, E., E. Gonzalez-Perez, and J. Modolell. 2007. "Apposition of iroquois expressing and non-expressing cells leads to cell sorting and fold formation in the Drosophila imaginal wing disc." *BMC Dev Biol* 7:106. doi: 1471-213X-7-106 [pii] 10.1186/1471-213X-7-106.

Vyas, N., A. Walvekar, D. Tate, V. Lakshmanan, D. Bansal, A. Lo Cicero, G. Raposo, D. Palakodeti, and J. Dhawan. 2014. "Vertebrate Hedgehog is secreted on two types of extracellular vesicles with different signaling properties." *Sci Rep* 4:7357. doi: 10.1038/srep07357.

Warren, N., and D. J. Price. 1997. "Roles of Pax-6 in murine diencephalic development." *Development* 124 (8):1573-82.

Watanabe, Y., and H. Nakamura. 2000. "Control of chick tectum territory along dorsoventral axis by Sonic hedgehog." *Development* 127 (5):1131-40.

Wheway, G., L. Nazlamova, and J. T. Hancock. 2018. "Signaling through the Primary Cilium." *Front Cell Dev Biol* 6:8. doi: 10.3389/fcell. 2018.00008.

Willaredt, M. A., K. Hasenpusch-Theil, H. A. Gardner, I. Kitanovic, V. C. Hirschfeld-Warneken, C. P. Gojak, K. Gorgas, C. L. Bradford, J. Spatz, S. Wolfl, T. Theil, and K. L. Tucker. 2008. "A crucial role for primary cilia in cortical morphogenesis." *J Neurosci* 28 (48):12887-900. doi: 10.1523/JNEUROSCI.2084-08.2008.

Willaredt, M. A., E. Tasouri, and K. L. Tucker. 2013. "Primary cilia and forebrain development." *Mech Dev* 130 (6-8):373-80. doi: 10.1016/j. mod.2012.10.003.

Wilson, S. W., and J. L. Rubenstein. 2000. "Induction and Dorsoventral Patterning of the Telencephalon." *Neuron* 28 (3):641-651.

Wilson, Stephen W., and Corinne Houart. 2004. "Review: Early Steps in the Development of the Forebrain." *Developmental Cell* 6:167-181. doi: 10.1016/S1534-5807(04)00027 9.

Woo, K., J. Shih, and S. E. Fraser. 1995. "Fate maps of the zebrafish embryo." *Curr Opin Genet Dev* 5 (4):439-43.

Xavier, G. M., M. Seppala, W. Barrell, A. A. Birjandi, F. Geoghegan, and M. T. Cobourne. 2016. "Hedgehog receptor function during craniofacial development." *Dev Biol* 415 (2):198-215. doi: 10.1016/j.ydbio.2016.02.009.

Yao, Y., P. J. Minor, Y. T. Zhao, Y. Jeong, A. M. Pani, A. N. King, O. Symmons, L. Gan, W. V. Cardoso, F. Spitz, C. J. Lowe, and D. J. Epstein. 2016. "Cis-regulatory architecture of a brain signaling center predates the origin of chordates." *Nat Genet* 48 (5):575-80. doi: 10.1038/ng.3542.

Zeltser, L. M. 2005. "Shh-dependent formation of the ZLI is opposed by signals from the dorsal diencephalon." *Development* 132 (9):2023-33.

Zeltser, L. M., C. W. Larsen, and A. Lumsden. 2001. "A new developmental compartment in the forebrain regulated by Lunatic fringe." *Nat Neurosci* 4 (7):683-4.

BIOGRAPHICAL SKETCH

Béatrice C. Durand

Affiliation: Sorbonne Université, Paris, France

Education:

1998-2002: Postdoctorate Baylor College of Medicine Houston - Texas, USA. Pr M. Jamrich

1994-1998: Postdoctorate University College London - London U.K. Pr M. Raff

1987-1993: Ph. D. thesis in Molecular and Cellular Biology - Universite Louis Pasteur Strasbourg, FR - Supervisor: Pr P. Chambon

1983-1986: Ecole Normale Supérieure Saint-Cloud. Agrégation de Biochimie Génie Génétique, First Rank.

Business Address:

Sorbonne Université CNRS UMR7622 – Laboratoire de Biologie du Développement, Bureau 714 - Campus Jussieu, Bâtiment C 7ème étage, boîte 24 9 Quai Saint Bernard 75252 Paris Cedex 05 France

Research and Professional Experience:

Scientific Leader, Administrative and Teaching responsibilities

Implementation of scientific projects from their conception to their development and publication, together with the supervision of the scientific, technical and management aspects with scientists, students and technical staff.

Advisor of PhDs (3), postdoctorates (2), students/technicians (16).

Expertise for scientific journals, grant agencies, PhD and HDR reports, recruitment committees.

Organisation of the "European Amphibian Club 2017 and 2019"

Editor for Journal of Developmental Biology

Fundings: LNCC (2019-2022) 45K€ - FPGG (2013) 150K€ -ARC Project (2009) 50K€- ANR Switch 300K€.

Teaching: Neuroscience Course of the Pasteur Institute "Development and Plasticity of the Nervous System." Teaching Assistant in the National School of Biotechnology in Strasbourg (E.N.S.B.S) - French National Education School teacher of

Professional Appointments:

2019-present: Sorbonne Université CNRS UMR7622 Developmental Biology Laboratory

2014-2019: Curie Institute, Orsay, FR

2007-2014: Institut de Biologie de l'Ecole Normale Superieure, Paris, FR

2003- 2006: Pasteur Institute, Paris, FR

Honors:

1998-2000: EMBO Fellow.

1994-1998: Human Research Frontier Fellow.

Publications from the Last 3 Years: (Sena et al. 2020) (Sena et al. 2019) (Sena, Feistel, and Durand 2016) (Durand 2016)

Durand, B. C. 2016. "Stem cell-like Xenopus Embryonic Explants to Study Early Neural Developmental Features In Vitro and In Vivo." *J Vis Exp* (108):e53474. doi: 10.3791/53474.

Sena, E., J. Bou-Rouphael, N. Rocques, C. Carron-Homo, and B. C. Durand. 2020. "Mcl1 protein levels and Caspase-7 executioner protease control axial organizer cells survival." *Dev Dyn*. doi: 10.1002/dvdy.169.

Sena, E., K. Feistel, and B. C. Durand. 2016. "An Evolutionarily Conserved Network Mediates Development of the zona limitans intrathalamica, a Sonic Hedgehog-Secreting Caudal Forebrain Signaling Center." *J. Dev. Biol.* 4:31. doi: 10.3390.

Sena, E., N. Rocques, C. Borday, H. S. Muhamad Amin, K. Parain, D. Sitbon, A. Chesneau, and B. C. Durand. 2019. "Barhl2 maintains T cell factors as repressors and thereby switches off the Wnt/beta-Catenin response driving Spemann organizer formation." *Development* 146 (10). doi: 10.1242/dev.173112.

In: The Forebrain
Editor: Morten F. Thorsen

ISBN: 978-1-53618-407-5
© 2020 Nova Science Publishers, Inc.

Chapter 2

CONTRIBUTION OF THE BASAL FOREBRAIN NUCLEI TO THE CONTROL OF CORTICAL ACTIVITY IN RODENTS

Irene Chaves-Coira[1], PhD,
Jesús Martín-Cortecero[1,2], PhD,
Margarita L Rodrigo-Angulo[1], PhD
and Angel Nuñez[1,*], PhD

[1]Department of Anatomía, Histología y Neurociencia. Universidad Autónoma de Madrid. Madrid. Spain

[2]Institute of Physiology and Pathophysiology, Heidelberg University, Heidelberg. Germany

ABSTRACT

Numerous evidences support that specific projections between different basal forebrain (BF) nuclei and their cortical targets are necessary to modulate cognitive functions in the cortex. These nuclei provide most of the cholinergic innervation to sensory, motor and prefrontal cortices,

---

[*] Corresponding Author's Email: angel.nunez@uam.es.

and hippocampus. Data suggest that these BF nuclei are integrated in distinct BF-cortical networks that may play different roles in sensory processing, motor control, or cortical arousal. The aim of this review is to show the existence of specific neuronal populations in the BF linking with specific sensory, motor and prefrontal cortices. In addition, electrophysiological properties of cholinergic pathways to control cortical activity are shown. Finally, the results obtained mainly in rodents could explain some aspects that are observed in neurodegenerative diseases.

1. INTRODUCTION

The basal forebrain (BF) has been implicated in a variety of behavioral functions, including learning, memory, attention and arousal (Buzsáki et al., 1988; Sarter and Bruno, 1997; Hasselmo 2006). It contains a heterogeneous population of neurons, including cholinergic and GABAergic projection neurons, as well as several interneurons (Zaborszky and Duque, 2000). The major target of BF projections is the cerebral cortex, providing the primary source of acetylcholine (ACh; Mesulam et al., 1983). Electrophysiological studies in the BF in combination with electroencephalographic (EEG) recordings indicate that cortical activation depends on BF inputs to the cortex (Metherate et al., 1988; Nuñez, 1996; Duque et al., 2000; Manns et al., 2000). Most of these effects have been explained by the release of ACh during wakefulness as well as in the rapid eye movement (REM) sleep (Celesia and Jasper, 1966; Jasper and Tessier, 1971; Rasmusson et al., 1992; Golmayo et al. 2003).

The cholinergic system also contributes to sensory and cognitive functions as shown in numerous behavioral studies and cognitive tasks and it has been implicated in attention sensory-coding, motivation, memory, and experience-dependent cortical plasticity (Hasselmo et al., 1995; Himmelheber et al., 2000; Rasmusson 2000; Hasselmo, 2006). Therefore, reduction of cholinergic activity can cause the generation or the aggravation of different neurodegenerative diseases such as Alzheimer, Schizophrenia or Parkinson disease. The aim of the present review is to analyze the role of the BF in the control of cortical activity in normal and pathological conditions.

2. ANATOMY

The BF is a non-well defined brain region located, as indicated by its name, in the mediobasal part of the mammalian telencephalon. Different authors have included distinct structures and nuclei in the BF, which are distributed along the rostro-caudal axis of the forebrain. It is formed of several heterogeneous structures which include the following nuclei: the medial septum, the horizontal and vertical limbs of the diagonal band of Broca (HDB and VDB, respectively), the substantia innominata, the extended amygdala, subpalial regions, and the nucleus basalis magnocellularis (B nucleus, or in humans, Meynert basal magnocellular nucleus; Mesulam et al., 1983; Nieuwenhuys et al., 2009).

The gross anatomy termed this structure as the anterior perforated substance because, to the naked eye, it appears to be perforated by many holes, which are actually blood vessels (Nieuwenhuys et al., 2009). The B nucleus is ventral to the globus pallidus and within an area known as the substantia innominata. The substantia innominata is a stratum in the human brain consisting partly of grey and partly of white substance, which lies below the anterior part of the thalamus and lentiform nucleus. The BF can be topographycally subdivided into a rostromedial, an intermediate and a caudolateral part. In the rostromedial part is situated the medial septal nucleus; the intermediate part co-extends roughly with the HDB and VDB, whereas the large caudolateral part is embedded in the substantia innominata and corresponds to the basal nucleus of Meynert (Nieuwenhuys et al., 2009).

These nuclei are conformed by a broad variety of neurons of different biochemical characteristics; they contain GABAergic, cholinergic and glutamatergic projecting neurons, as well as interneurons. These structures provide most of the cholinergic innervation to sensory, motor and prefrontal cortices and the hippocampus (Houser et al., 1985; Mesulam et al., 1992; Gritti et al., 1997, 2003; Bloem et al., 2014; Zaborszky et al., 2015; Kim et al., 2016).

From a biochemical point of view based on the classification of Mesulam (1983) we can make a classification of cholinergic neurons in 6 groups. BF contains six groups of cholinergic neurons, designated Ch1-Ch6, based on cytoarchitectonic criteria and connectivity patterns. In the rodent brain, the Ch1-Ch2 sectors are contained within the medial septum and VDB, respectively, providing the main cholinergic projection to hippocampus formation. The Ch3 sector is mostly contained within the lateral side of the HDB and provides the main cholinergic projection to the olfactory bulb. The Ch4 sector contains the cholinergic neurons of the basal nucleus of Meynert or B nucleus, the substantia innominata and probably also neurons located laterally of VDB. One criterion for unification of Ch4 is that its components provide the main cholinergic innervation to the neocortex and to the amygdala. The Ch5-Ch6 sectors are located in the pontomesencephalic reticular formation and provide the main cholinergic innervation to the thalamus. The cortical mantle is under a dual cholinergic influence, in part because there is overlap in connectivity whereby individual sectors of Ch. On the one hand, a monosynaptic corticopetal cholinergic pathway comes mainly from Ch4. On the other hand, there is an indirect pathway from Ch5-Ch6 to the thalamus, which modulate cortical activity.

From an anatomical point of view, one broad accepted classification of BF pathways to the cortex is the one that Kristt et al., did in 1985, where they divided the BF fibers in 3 different pathways (Kristt, et al. 1985):

1) Anterior pathway: fiber bundles can be traced from the substantia innominata and B nucleus to the frontal cortex.
2) Medial pathway: The existence of a medial pathway rich in cholinergic fibers, histologically described by Shute and Lewis (1975). Emerging from the HDB is a bundle of fibres with a dorso-lateral course and entering layer VI of the anterior medial cortex in rodents and humans.
3) Lateral pathway: Seems to innervate the frontal-lateral, parieto-temporal and latero-occipital cortex, mainly emerging from the substantia innominata and the basal nucleus of Meynert (or B nucleus).

Early anatomical descriptions of cholinergic projections were consistent with the notion of a diffuse pathway from the BF to the cortex (Descarries et al., 1987). Nearly all cortical areas are innervated by BF cholinergic neurons (Lysakowski et al., 1989). For a long time, the prevailing view was that cholinergic neurons in the BF release ACh into the cortex nonspecifically and that ACh acts via volume-transmission to regulate the excitability of cortical neurons (Descarries et al., 1997). However, the traditional description of the forebrain cholinergic system as a diffusely organized neuromodulator system has been replaced by a more precise distribution by which the cholinergic system from the BF may be innervating the neocortex at a spatially more refined scale than previously considered. Recent anatomical studies have indicated the existence of a highly structured and topographic organization of BF efferent projections to sensory cortices (Zaborszky, 2002; Zaborszky et al., 2005, 2015).

Studies in rodents of anatomical pathways linking the BF with cortical areas have shown that separate or partially overlapping groups of BF neurons display specific projection pathways to primary sensory cortices of different modalities (Zaborsky et al., 2015; Chaves-Coira et al., 2016). The HDB/VDB area is mainly related with primary sensory cortices with specific projections to somatosensory, visual or auditory primary cortices according to the rostro-caudal neuronal distribution. In contrast, B nucleus projects to all sensory cortical areas without a clear specific pattern. Moreover, differences between HDB/VDB and B nuclei are also evident in their projections to the medial prefrontal cortex (mPFC). HDB/VDB, but no B nucleus, shows bidirectional projections to prelimbic and infralimbic areas of the mPFC. B nucleus projects to the secondary motor cortex while the projection from HDB/VDB to the secondary motor cortex is less important (Figure 1). Present data suggest that these BF areas are integrated in distinct BF-cortical networks that may play different roles in sensory processing, motor control, or cortical arousal (Chaves-Coira et al. 2018a).

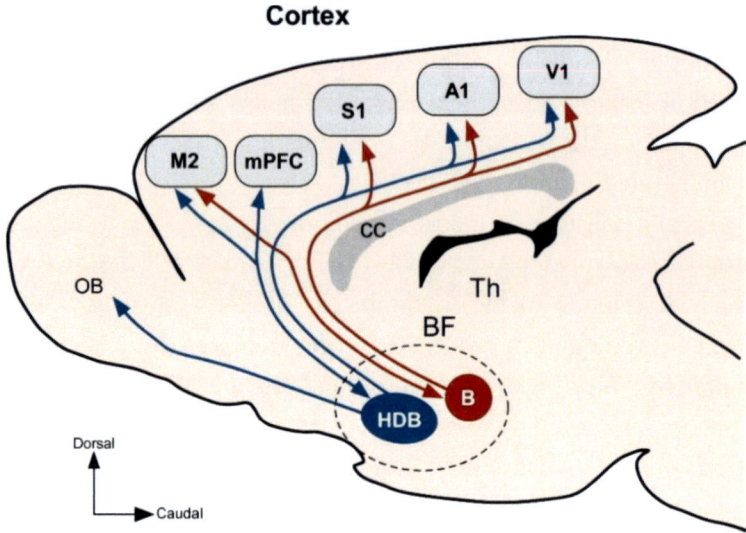

Figure 1. Summary diagram displaying anatomical connections from the BF to the cortex. Results showed that HDB has preferential projections to sensory cortices and mPFC cortices while the B nucleus projects to most of cortical areas. Modified from Chaves-Coira et al. 2018a.

Application of retrograde tracers in both hemispheres of the primary somatosensory, auditory or visual cortical areas have shown labeled neurons in the ipsi- and contralateral areas of the diagonal band of Broca and substantia innominata, indicating that these projections are bilateral. In contrast, the B nucleus only showed ipsilateral projections to the cortex (Chaves-Coira et al. 2018b).

It is well stablished that the BF cholinergic innervation of the mPFC is crucial for cognitive performance (e.g., Sarter and Bruno, 1997; Dalley et al., 2004; Parikh et al., 2007; Hasselmo and Sarter, 2011; Howe et al., 2013). Using retrograde tracers, BF shows that neurons projecting to the mPFC tend to cluster in the rostral portion of the BF, which also receive projections from the mPFC (Chandler et al., 2013; Bloem et al., 2014; Chaves-Coira et al., 2018a).

Cholinergic afferents to the cortex are distributed at high density throughout all layers of the neocortex in rodents, with particularly high densities in cortical layers 1, 5 and 6 (Radnikow and Feldmeyer, 2018). In the human neocortex, highest density of cholinergic receptors is observed in superficial layers of most cortical areas (Obermayer et al., 2017).

Consistent with these anatomical findings, optogenetic activation of cholinergic neurons in BF induces desynchronization in specific sensory cortices and sensory modulation (Chaves-Coira et al., 2016, 2018a; Kim et al., 2016). Optogenetic stimulation directed towards the HDB facilitated sensory responses in the primary somatosensory cortex of rodents through activation of muscarinic receptors. The same optogenetic stimulation of HDB induced an inhibition of sensory responses in mPFC by activation of nicotinic receptors (Chaves-Coira et al., 2018a), indicating the specificity of the cholinergic projections to the cortex (see below).

3. CHOLINERGIC RECEPTORS

The existence of different types of cholinergic receptors with a heterogeneous location in the cortex strongly suggest that ACh acts in a specific manner. Cholinergic receptors are composed of two classes of transmembrane macromolecular complexes, the muscarinic receptor (mAChR) and the nicotinic receptor (nAChR) families, each of which is further divided into subclasses, which allows the ACh to modulate cortical activity in very different ways (McCormick, 1992). The nAChR are pentameric ionotropic receptors that exert fast neuronal actions (Dani and Bertrand, 2007; Bloem et al., 2014). The mAChRs are metabotropic receptors activating heterotrimeric G-proteins and acting through a cascade of intracellular reactions that leads to modulate K^+ and Ca^{2+} channels (Bubser et al., 2012; Dasari et al., 2017). The effect produced by activation of mAChRs is slow and long lasting, compared to that generated by nAChR activation, which is rapid and short lasting (Gulledge et al., 2007; Bloem et al., 2014). The pre-synaptic mAChRs M2 and M4 preferentially couple to Gi and Go proteins that generally have inhibitory effects on voltage-activated Ca^{2+} channels or open K^+ channels, resulting in a decrease of neurotransmitter release (Qian and Saggau, 1997; Fernandez de Sevilla and

Buño, 2003). M1, M3 and M5 subtypes are preferentially coupled to Gq and G11 proteins and are mainly located in the postsynaptic neuron.

The expression of the different types of receptors is also heterogeneous. In the rodent cortex, mAChRs family mainly M1, M2 and M4 are expressed in the neocortex, although M4 has a considerably lower expression. M1-4 mAChRs are expressed in the hippocampus. The immune-reactive staining of mAChRs shows a strong laminar pattern (Levey et al., 1991). M1 mAChR immuno-reactivity is present in most cortical neurons and particularly dense in layer 2/3 and layer 6. M2 protein is dense in layer 4 and in the border of layer 5/6. M4 mAChR immunoreactivity is localized in layer 2/3, layer 4 and layer 5. In the human neocortex, highest densities of M1, M2 and M3 mAChRs are observed in superficial layers of the cortex (Obermayer et al., 2017).

The distribution of nAChRs is also heterogeneous in the cortex. The pyramidal cells of layer 2/3 practically do not express nAChRs, although cholinergic modulation may exist throughout interneurons. Thus, the application of ACh in layers 2/3 produce an inhibition of the response of pyramidal neurons mediated by nicotinic receptor activation expressed in interneurons (Bloem et al., 2014; Verhoog et al., 2016). However, the pyramidal neurons of layer 5 and layer 6 are directly modulated by nicotine receptors (Gulledge et al., 2007; Albuquerque et al., 2009; Bloem et al., 2014; Verhoog et al., 2016), so that, the ACh can modulate neuronal plasticity in a specific way in each cortical layer.

In addition to the complex effect that cholinergic neurons exert in the cortex there is another cell type population in the BF that may play an important role in the regulation of cortical activity, the cortically-projecting GABAergic neurons. It has been identified different subtypes of GABAergic neurons such as calbindin, calretinin and parvalbumin expressing GABAergic neurons that projects to the cortex mainly targeting cortical GABAergic interneurons (Freund and Meskenaite, 1992; Duque et al., 2000). Thus, the effect of GABAergic projecting neurons is a disinhibition of pyramidal cells that widely contributes to arousal and the generation of fast rhythms (Jimenez-Capdeville et al., 1997), increasing the fidelity of sensory processing and attention processes.

4. CHOLINERGIC ACTIVATION OF CORTICAL ACTIVITY

Moruzzi and Magoun (1949) were the first authors in demonstrate that cerebral activation is related to changes in EEG waves by means of the brainstem reticular arousal system. The more evident arousal effect on EEG activity is the "desynchronization" phenomenon. It refers to the rapid shift from high-amplitude low-frequency waves in the EEG, typical of the slow wave sleep (SWS) and anesthesia, to low-amplitude high-frequency cortical activity, typical of wakefulness and REM sleep. Later, it has been shown that cholinergic modulation of the cortical EEG activation and its functional correlates (arousal or attention) is exerted via anatomical projections from the BF (Celesia and Jasper, 1966; Jasper and Tessier 1971; Buzsaki et al., 1988; McCormick et al., 2015). ACh release is maximal during wakefulness and REM sleep, while it decreases during SWS (Jasper and Tessier, 1971; Marrosu et al., 1995; Arrigoni et al., 2010).

Unit recordings in the BF have indicated that most cells have higher firing rate during fast cortical activity (Nuñez, 1996; Dringenberg and Vanderwolf, 1997; Detari et al., 1999; Manns et al., 2000a). As indicated above, this neuronal population includes cholinergic neurons, GABAergic neurons and unidentified neurons that increased their firing preceding changes of the EEG (Duque et al., 2000). It has been described two types of cholinergic neurons: "Early firing neurons" (~70%), which are more excitable and "Late firing neurons" (~30%), which are less excitable. These two cholinergic cell populations might be involved in distinct functions: the early firing group being more suitable for phasic changes in cortical acetylcholine release associated with attention while the late firing neurons could support general arousal by maintaining tonic ACh levels (Unal et al., 2012).

Since GABAergic BF axons were found to terminate exclusively on cortical GABAergic interneurons (Freund and Meskenaite, 1992), these findings are compatible with the notion that at least a subpopulation of PV-containing BF neurons promotes functional activation in the cerebral cortex by disinhibition (Jimenez-Capdeville et al., 1997; Duque et al., 2000). Besides to neurons that increased their firing during cortical activation, several studies described the presence of a smaller number of BF neurons that reduced their firing during EEG activation (Nuñez, 1996; Manns et al., 2000b). These neurons are activated antidromically from the cortex and it was suggested that they may be GABAergic neurons inhibiting cholinergic and/or GABAergic cells (Detari et al., 1997).

Recently, it has been suggested that cholinergic BF neurons may induce a complex effect in the cortex by co-release of ACh and GABA neurotransmitters. Saunders et al. (2015), for the first time, provide direct and compelling evidences about ACh and GABA co-release in cholinergic neurons of the basal forebrain. A population of cholinergic cells can produce the enzyme to synthesize GABA as well as the protein to pack GABA into vesicles. In this study they demonstrate by means of optogenetic stimulation that this BF neurons can trigger postsynaptic potentials in layer 1 cortical interneurons, mediated by the activation of $GABA_A$ and nAChRs. In addition, Takács et al. (2018) have demonstrated that BF may exert a dual effect in the hippocampal cells, inducing inhibition followed by excitation due to activation medial septum neurons that co-release both ACh and GABA neurotransmitters. Consequently, this feature enriches and add complexity about the functional role of basal forebrain.

BF has been also indicated as an important structure in the generation of SWS. Electrical stimulation of the HDB and lateral preoptic area evokes SWS while lesions of these areas suppress SWS and REM sleep in cats (Madoz and Reinoso-Suarez, 1968; McGinty and Sterman, 1968; Szymusiak and McGinty, 1986). In addition, Szymusiak and McGinty described in cats some BF cells in cat that increased their discharge in anticipation of NREM sleep onset. These "sleep active" neurons were antidromically driven from the external capsule and cingulate bundle and tentatively identified as either cholinergic or GABAergic neurons (Szymusiak and McGinty, 1986; McGinty and Szymusiak, 2000).

Cholinergic projections from the BF also regulates rhythmic activity in the hippocampus and neocortex. Hippocampal theta oscillations are large rhythmic neural activities in the hippocampus and many other hippocampus-associated brain regions. Theta rhythm originates in the medial septum-diagonal band of Broca (Gaztelu and Buño, 1982). This rhythm can be divided in type 1 movement related theta (>8 Hz) and type 2 sensory processing theta (4-8 Hz; Vertes and Kocsis 1997; Buzsaki, 2002; Bland, 2009). The type 2 theta rhythm is dependent on ACh inputs from the medial septum-diagonal band of Broca and has been associated to arousal, alertness and sensory processing (Vertes, 2005; Hasselmo, 2006; Buzsáki, 2002). In agreement with that, ACh release occurs over many seconds after the appearance of spontaneous or induced theta oscillations in urethane-anesthetized rats and optogenetic activation of cholinergic medial septal neurons induces hippocampal theta oscillations (Zhang et al., 2010). It was further demonstrated that M1 muscarinic receptors in pyramidal neurons but not interneurons play a critical role in cholinergic modulation of hippocampal synaptic plasticity and theta generation (Gu et al., 2017). These results were further substantiated by *in vivo* observations of reduced theta power and impaired Y-Maze performance in mice with selective receptor knockout M1 in pyramidal neurons (Gu et al., 2017). Other cholinergic receptors, such as nicotinic or muscarinic M4 receptors, may have also contributed to cholinergic-dependent theta rhythm generation (Stoiljkovic et al., 2016; Gu et al., 2017). Intracellular recordings during hippocampal theta *in vivo* show that the membrane potential of hippocampal neurons oscillates at theta frequencies. This "intracellular theta" rides on a sustained, cholinergic-dependent depolarization and consists in rhythmic EPSPs (Nuñez et al., 1987; Bland et al., 2002) and IPSPs (Hangya et al., 2009) evoked by rhythmic firing of medial septum and diagonal band of Broca neurons (Nuñez et al., 1990).

The cholinergic facilitatory effect of EEG activation is also observed in the generation of fast rhythms. Gamma oscillations are synchronous network oscillations in the 30–100 Hz range found throughout the neocortex and hippocampus. Rhythms in the gamma range can establish synchronization of distributed neural responses throughout the brain (Singer, 1999). The entrainment of neuronal activity to this high-frequency oscillation is thought to be important for the timing of spikes both within and between different brain structures determining the flow of information and facilitating the information processing (Ainsworth et al., 2012; Betterton et al., 2017). The occurrence of gamma oscillations is a critical feature of attention and information processing (Fries et al., 2001; Womelsdorf et al., 2006) and underlies the synaptic plasticity required for the encoding of long-term memory (Buzsáki and Draguhn, 2004; Buzsaki and Wang, 2012). The release of ACh in the neocortex and hippocampus activates nAChRs and mAChRs that regulate the processing of information within these circuits (Hasselmo, 2006). In agreement with that, cholinergic agonists and acetylcholinesterase inhibitors can induce gamma oscillations in the 30–100 Hz frequency range in the neocortex and hippocampus *in vitro* (Spencer et al., 2010, Betterton et al., 2017) and *in vivo* (Cape et al., 2000, Rodriguez et al., 2004). In the hippocampus, gamma oscillations can coincide with theta oscillations (Buzsaki et al., 1992, Lisman and Jensen, 2013). Consequently, ACh may facilitate the generation of both type of oscillations in the cortex, facilitating the performance of learning, memory or attention tasks by means of an increase of neuronal synchronization (Howe et al., 2017).

5. CHOLINERGIC MODULATION OF SENSORY RESPONSES

The functional role of Ach as a neuromodulator that increases the excitability of the entire cortical mantle, and promotes the information processing in the awake animal and during REM sleep, has been widely demonstrated, through a wide range of stimuli, behavioral manipulations, novelty, sensory stimulation, which they all increase cholinergic activity in the brain. Most of these effects in sensory information processing are mediated by mAChRs, which affects are long-lasting. The following cholinergic effects observed in the cortex explain these roles: (i) high ACh levels increase the magnitude of afferent input through activation of nAChRs located presynaptically on thalamocortical terminals (Disney et al., 2007); (ii) high ACh levels acting through mAChRs reduce cortico-cortical excitatory recurrent interactions through presynaptic inhibition of glutamate release (Hasselmo and Bower, 1992, Eggermann and Feldmeyer, 2009); (iii) stimulation of mAChRs depolarizes and increases neuronal excitability and responsiveness for several minutes (Metherate et al., 1988, Oldford and Castro-Alamancos, 2003); (iv) stimulation of mAChRs reduces spike frequency adaptation by inhibiting the K+-mediated M-current and the slow afterhyperpolarizing current (McCormick and Prince, 1986, Hasselmo, 2006). Consequently, the mAChR-mediated depolarization is the main responsible to boost neuronal responses and increase the signal-to noise ratio in the sensory cortex (Sillito and Kemp 1983; Metherate et al., 1988; Pinto et al., 2013; Minces et al., 2017), thereby enhancing neuronal response reliability.

Furthermore, the long-term modifications in synaptic efficacy induced by ACh have been widely proposed to be the cellular basis of the learning machinery of the brain and in attentional processes (Hasselmo and Sarter, 2011; Nabavi et al., 2014; Gruart et al., 2015). For example, cholinergic axons in the barrel cortex of rodents increase activity during whisking (Eggermann et al., 2014). Herrero et al. (2008) showed that in the primary visual cortex, blockade of mAChRs but not nAChRs impaired the attentional modulation of primary visual cortical neurons. Thanks to the development of choline oxidase microelectrode technique that provide a high resolution for ACh detection (Parikh et al., 2004), it has been demonstrated that there are a rapid transient increase of ACh during attentional tests that would facilitate responses to relevant stimuli (Dalley et al., 2004; Parikh et al., 2007; Howe et al., 2013), contrary to the traditional criteria that considered more important the slow and tonic increase of excitability evoked by the ACh.

The evoked ACh release in the cortex was modality-specific. For example, visual stimulation evoked a larger ACh increase in the visual cortex than in the somatosensory cortex whereas skin stimulation had the opposite effect (Golmayo et al., 2003; Fournier et al., 2004). In the somatosensory cortex, ACh facilitates the response to whisker stimulation in anesthetized rats and unmasked a receptive field that was not present before ACh was applied (Donoghue and Carroll, 1987; Metherate et al., 1987; Rasmusson, 2000). The enhanced response lasted as long as 20 min after ACh application. BF stimulation triggers an atropine-sensitive enhancement of responses evoked by vibrissa deflection in layer 5 neurons that is mainly due to an enhanced NMDA-mediated response (Nuñez et al., 2012, Barros-Zulaica et al., 2014, Chaves-Coira et al., 2018). The strong

Ca^{2+} signal associated with NMDA receptor activation triggers LTP, increasing sensory detection and processing (Alenda and Nuñez, 2007, Hasselmo and Sarter, 2011, Barros-Zulaica et al., 2014, Nabavi et al., 2014) and regulates whisking (de Kock and Sakmann, 2009).

ACh also regulates information processing in the primary visual cortex not only by regulating the magnitude of the visual response, but also the selectivity to stimulus-features such as orientation, direction, and size (Herrero et al., 2008; Soma et al., 2013). Repetitive electrical stimulation of fibers from the lateral geniculate nucleus to the primary visual cortex induces LTP in the cortex that is enhanced by stimulation of the BF area through activation of mAChRs (Dringenberg et al., 2007). ACh also increases the auditory responses of cortical neurons in a non-specific manner (Jimenez-Capdeville and Dykes, 1996). Moreover, learning-induced cortical plasticity is augmented by cortical application of ACh and conversely, prevented by cortical application of the muscarinic antagonist, atropine, during the conditioning (Ji and Suga, 2003, Chen and Yan, 2007).

Martin Sarter's group have suggested that ACh would not only be modulating the cognitive processes through a tonic response activity but also through a fast-phasic increase in ACh concentration when the animal must perform attentional tests (Sarter et al., 2009). This fast cholinergic activity have been observed in the mPFC (activate in the range for 200 miliseconds and last for few seconds) in attentional tests when the cue is detected (Parikh et al., 2007; Parikh and Sarter, 2008; Howe et al., 2013). However, these rapid transient increases are absent in the motor cortex in studies with the animal moving, suggesting that the fast ACh transients are released in the mPFC in tasks that demand attentional efforts and with cue detection (Parikh et al., 2007; Howe et al., 2013).

6. CHOLINERGIC MODULATION OF PREFRONTAL CORTEX

In both humans and rodents, the mPFC plays a key role in many higher executive functions (Groenewegen and Uylings, 2000; Miller and Cohen, 2001; Dalley et al., 2004). Results obtained from retrograde tracer injections in the dorsal and ventral regions of the mPFC indicated that sensory inputs arrives at the dorsal mPFC through secondary sensory cortical areas, and through the insular and temporal cortical areas, at least in rodents (Martin-Cortecero and Nuñez, 2016). Authors suggest that mPFC plays an important role in sensory processing, which may have important implications in attentional and memory processes. It is therefore reasonable to think that ACh should modulate this cortical area.

The bidirectional connectivity between mPFC and BF has attracted great interest as a circuit involved in modulating decision making, cortical arousal, and learning and memory processes (Groenewegen and Uylings, 2000; Miller and Cohen, 2001; Dalley et al., 2004; Bloem et al., 2014; Zaborszky et al., 2015). Since previous data have described cholinergic modulation in sensory cortices, the modulating effect of ACh on the prefrontal cortex should be similar. In fact, there are not large differences in nAChRs and mAChRs location between different cortical areas (Clarke et al., 1984). In addition to this, it has been shown that layer 5 neurons of the mPFC are modulated mainly by M1 muscarinic receptors (Gulledge et al., 2009), as it is also described, for example, in the somatosensory cortex (Nuñez et al., 2012); in contrast, in layer 2/3 and layer 6 neurons the cholinergic projection is minor (Houser et al., 1985).

In mPFC ACh appears to increase the inhibitory activity of neurons located in superficial layers while seems to exert a facilitatory effect on the thalamic incoming information (Verhoog et al., 2016). In other words, it has been propose that there would be a facilitation of the incoming information from the thalamus, while inhibiting at the same time much of the associational cortico-cortical information that comes mostly to superficial layers of the mPFC, which would facilitate the thalamic incoming information to be processed more accurately; it would also facilitate the downstream information from 5/6 layers of the mPFC to subcortical structures, such as the striatum, affecting in this way the onset of the motor response, "go activity", and therefore orienting and influencing directly in the execution of the response (Sarter et al., 2005; Sarter et al., 2014; Bloem et al., 2014; Verhoog et al., 2016), and serving as gate control and sensory filtering of ascending information through the reticular thalamic nucleus and

Zona Inncerta (Wimmer et al., 2015; Escudero and Nuñez, 2019; Nakajima et al., 2019).

The connections between the mPFC and other cortical areas that apparently contribute to information processing seem to be specific topographically organized (Hoover and Vertes, 2007; Zaborszky et al., 2015). In such processes, the cholinergic, GABAergic and glutamatergic BF neuronal groups or subgroups probably play diverse and complementary roles. Most of projecting neurons are mainly concentrated in the rostral BF, including VDB and HDB (Chaves-Coira et al., 2018). Their activation by optogenetic stimulation induced an important facilitation of whisker responses in the primary somatosensory cortex through activation of mAChRs (Chaves-Coira et al., 2016). However, the same BF projection inhibited whisker responses in mPFC though activation of nAChRs (Chaves-Coira et al., 2018). As indicated above, this effect may be due to the inhibition exerted by ACh on the associational cortico-cortical inputs mostly to superficial layers of the mPFC since sensory inputs to the mPFC arise from cortico-cortical projections.

Furthermore, transient, or phasic, increases in ACh release in PFC mediate cue detection (Parikh et al., 2007; Howe et al., 2013; Gritton et al., 2016). Cortical pyramidal cells, GABAergic interneurons, and the terminals of thalamic afferent projections all express cholinergic receptors (Disney and Aoki, 2008; Parikh et al., 2008), giving the cortical cholinergic system the capacity to shape activity that emerges from interactions between cortical neurons during attentional performance. This idea is supported by evidence indicating that stimulus-evoked spike coherence is sensitive to manipulations of cholinergic activity (Rodriguez et al., 2004; Herrero et al., 2008), and that large-scale synchronous neural activity induced by attended stimuli requires the activation of mAChRs (Rodriguez et al., 2004). Furthermore, it has been shown that theta oscillations accompany behaviorally relevant stimuli (see above), and that neurons of the BF support the generation of both theta and gamma oscillations in the rodent PFC (Howe et al., 2017).

7. BASAL FOREBRAIN AND PATHOLOGY

The disruption of cholinergic system and its participation in cognitive impairment is an old issue (Fibiger, 1991). The degeneration of the BF is associated with Alzheimer's disease (AD; Whitehouse et al., 1982; Grothe et al., 2012; Hampel et al., 2018), different forms of dementia (Cummings and Benson, 1984) and with normal cognitive aging (Gallagher and Colombo, 1995). Although AD is most classically associated with memory deficits, these deficits are typically conflated with attentional issues in which cholinergic BF system is clearly involved (Romberg et al., 2013). The cholinergic hypothesis is supported from evidences of loss of cholinergic markers in the cortex (Geula and Mesulam, 1989), loss of the number of neurons in the BF (Whitehouse et al., 1982), and the report of a volume loss in the nucleus basalis of Meynert in mild cognitive impairment patients (prodromal signs of AD; Cantero et al., 2016).

There is a deficiency of choline acetyl transferase (ChAT) in the neocortex, amygdala, and hippocampus of AD patients, revealing that cortical regions containing the greatest density of neurofibrillary tangles exhibited the maximum reduction of ChAT (Davies and Maloney 1976). Moreover, recent experiments have shown that individuals with subjective cognitive decline, a risk population for preclinical AD, exhibit smaller volume of Ch1–Ch2 and Ch4 subdivisions of the BF (Cantero et al., 2019; Scheef et al., 2019). Most studies show a loss of nAChRs in the cerebral cortex and a decrease of postsynaptic nAChRs on cortical neurons (Nordberg and Winblad, 1986; Schroder et al., 1991). With respect to mAChRs of the cerebral cortex, the M2 receptors are decreased and the M1 receptors of the cerebral cortex may be dysfunctional (Mash et al., 1985; Jiang et al., 2014).

In early work on cognitive decline in Parkinson's disease, autopsy studies identified degeneration of the cholinergic BF, in particular the nucleus basalis of Meynert, suggesting that cortical cholinergic depletion may be the major determinants of cognitive complications. This represents a neuropathological feature of Parkinson's disease shares with Alzheimer's disease dementia (Candy et al., 1983; Liu et al., 2015). In Parkinson's disease dementia, BF degeneration has been detected as volumetric reduction of the substantia innominata (Choi et al., 2012) (Lee et al., 2014) and in the nucleus basalis of Meynert (Ray et al., 2018).

Abnormal cognitive mechanisms that contribute to the development of psychotic symptoms are also attributed specifically to the aberrations in cortical cholinergic transmission. Experimental evidence from studies demonstrating amphetamine-induced sensitization of cortical cholinergic transmission as well as the ability of antipsychotic drugs to normalize the activity of cortical cholinergic inputs (Sarter et al., 2005).

CONCLUSION

The BF and mainly the cholinergic system are implicated in a variety of behavioral functions, including learning, memory, attention and arousal. To modulate these complex behaviors there are specific neuronal networks between the BF and the cortex that play different roles in the control of cortical activity. The existence of different types of cholinergic receptors with a heterogeneous location in the cerebral cortex indicate that ACh acts in a specific manner. The results reviewed here demonstrate a wide variety of cortical cholinergic projections and cholinergic receptor that allow complex control of cortical activities.

ACKNOWLEDGMENTS

Our work has been supported by the Spanish Ministerio de Economía y Competitividad Grants (BFU2012-36107 and SAF2016-76462 AEI/FEDER).

REFERENCES

Ainsworth, M., Lee, S., Cunningham, M. O., Traub, R. D., Kopell, N. J. and Whittington, M. A. (2012). Rates and rhythms: a synergistic view of frequency and temporal coding in neuronal networks. *Neuron*, 75:572–583.

Arrigoni, E., Mochizuki, T. and Scammell TE. (2010). Activation of the basal forebrain by the orexin/hypocretin neurones. *Acta Physiologica*, 198: 223–235.

Albuquerque, E. X., Pereira, E. F., Alkondon, M. and Rogers, S. W. (2009). Mammalian nicotinic acetylcholine receptors: from structure to function. *Physiological Review*, 89:73–120.

Alenda, A. and Nuñez, A. (2007). Cholinergic modulation of sensory interference in rat primary somatosensory cortical neurons. *Brain Research*, 1133:158–167.

Barros-Zulaica, N., Castejon, C. and Nuñez, A. (2014). Frequency-specific response facilitation of supra and infragranular barrel cortical neurons depends on NMDA receptor activation in rats. *Neuroscience*, 281:178-194.

Betterton, R. T., Broad, L. M., Tsaneva-Atanasova, K. and Mellor, J. R. (2017). Acetylcholine modulates gamma frequency oscillations in the hippocampus by activation of muscarinic M1 receptors. *European Journal of Neuroscience*, 45:1570–1585.

Bland, B. H., Konopacki, J. and Dyck, R. H. (2002). Relationship Between Membrane Potential Oscillations and Rhythmic Discharges in Identified Hippocampal Theta-Related Cells. *Journal Neurophysio-logy*, 88:3046–3066; doi:10.1152/jn.00315.2002.

Bloem, B., Poorthuis, R. B. and Mansvelder, H. D. (2014). Cholinergic modulation of the medial prefrontal cortex: the role of nicotinic receptors in attention and regulation of neuronal activity. *Frontiers Neural Circuits*, 8:17. doi: 10.3389/fncir.2014.00017.

Bubser, M., Byun, N., Wood, M. R. and Jones, C. K. (2012). Muscarinic receptor pharmacology and circuitry for the modulation of cognition.

Handbook Experimental Pharmacology, 208:121–166. doi: 10.1007/978-3-642-23274-9.

Buzsáki, G., Bickford, R. G., Ponomareff, G., Thal, L. J., Mandel, R., and Gage, F. H. (1988). Nucleus basalis and thalamic control of neocortical activity in the freely moving rat. *Neuroscience*, 8:4007–4026.

Buzsaki, G., Horvath, Z., Urioste, R., Hetke, J. and Wise, K. (1992). High-frequency network oscillation in the hippocampus. *Science,* 256:1025-1027.

Buzsaki, G. (2002). Theta oscillations in the hippocampus. *Neuron*, 33: 325-340.

Buzsáki, G., and Draguhn, A. (2004). Neuronal oscillations in cortical networks. *Science*, 304:1926-1929.

Buzsaki, G. and Wang, X. J. (2012). Mechanisms of gamma oscillations. *Annual Review Neuroscience*, 35:203–225.

Candy, J. M,, Perry, R. H., Perry, E. K., Irving, D., Blessed, G., Fairbairn, A. F, et al. (1983). Pathological changes in the nucleus of Meynert in Alzheimer's and Parkinson's diseases. *Journal Neurology Science*, 59: 277–89.

Cantero, J. L., Zaborszky, L., and Atienza, M. (2016). Volume loss of thenucleus basalis of Meynert is associated with atrophy of innervated regions in mild cognitive impairment. *Cerebral Cortex*, 27:3881–3889. doi: 10.1093/cercor/bhw195.

Cantero, J. L., Atienza, M., Lage, C., Zaborszky, L., Vilaplana, E., Lopez-Garcia, S., Pozueta, A., et al.; Alzheimer's Disease Neuroimaging Initiative. (2019). Atrophy of Basal Forebrain Initiates with Tau Pathology in Individuals at Risk for Alzheimer's Disease. *Cerebral Cortex,* pii: bhz224. doi: 10.1093/cercor/bhz224.

Cape, E. G., Manns, I. D., Alonso, A., Beaudet, A., Jones, B. E. (2000). Neurotensin-induced bursting of cholinergic basal forebrain neurons promotes gamma and theta cortical activity together with waking and paradoxical sleep. *Journal Neuroscience*, 20:8452-8461.

Celesia, G. G. and Jasper, H. H. (1966). Acetylcholine released from cerebral cortex in relation to state of activation. *Neurology*, 16:1053–1064.

Chandler, D. J., Lamperski, C. S., and Waterhouse, B. D. (2013). Identification and distribution of projections form monoaminergic and cholinergic nuclei to functionally differentiated subregions of prefrontal cortex. *Brain Research*, 1522:38-58. doi: 10.1016/j.brainres.2013.04.0575.

Chaves-Coira, I., Barros-Zulaica, N., Rodrigo-Angulo, M. L., and Nuñez, A. (2016). Modulation of specific sensory cortical areas by segregated basal forebrain7cholinergic neurons demonstrated by neuronal tracing and optogenetic stimulation in8mice. *Frontiers Neural Circuits*, 10:1-13. doi: 10.3389/fncir.2016.00028.

Chaves-Coira, I., Martin-Cortecero, J., Nuñez, A. and Rodrigo-Angulo, M. L. (2018a). Basal Forebrain Nuclei Display Distinct Projecting Pathways and Functional Circuits to Sensory Primary and Prefrontal Cortices in the Rat. *Frontiers in Neuroanatomy*, 12:69.

Chaves-Coira, I., Rodrigo-Angulo, M. L., and Nuñez, A. (2016). Bilateral Pathways from the basal forebrain to sensory cortices may contribute to synchronous sensory processing. *Frontiers in Neuroanatomy*, 12:5. doi: 10.3389/fnana.2018.00005.

Chen, L. and Yan, J. (2007). Cholinergic modulation incorporated with a tone presentation induces frequency-specific threshold decreases in the auditory cortex of the mouse. *European Journal Neuroscience*, 25: 1793–1803. doi: 10.1111/j.1460-9568.2007.05432.x.

Choi, S. H., Jung, T. M., Lee, J. E., Lee, S. K., Sohn, Y. H. and Lee, P. H. (2012). Volumetric analysis of the substantia innominata in patients with Parkinson's disease according to cognitive status. *Neurobiology of Aging*, 33: 1265–72.

Clarke, P. B., Pert, C. B. and Pert, A. (1984). Autoradiographic distribution of nicotinic receptors in rat brain. *Brain Research*, 223:390-395.

Dalley, J. W., Cardinal, R. N. and Robbins, T. W. (2004). Prefrontal executive and cognitive functions in rodents: neural and neurochemical substrates. *Neuroscience Biobehavior Review*, 28:771–784.

Dani, J. A. and Bertrand, D. (2007). Nicotinic acetylcholine receptors and nicotinic cholinergic mechanisms of the central nervous system. *Annual Review Pharmacology Toxicology* 47:699-729.

Dasari, S., Hill, C., and Gulledge, A. T. (2017). A unifying hypothesis for M1 muscarinic receptor signalling in pyramidal neurons. *Journal Physiology (London)* 595:1711–1723. doi: 10.1113/jp273627.

Davies, P., Maloney, A. J. (1976). Selective loss of central cholinergic neurons in Alzheimer's disease. *Lancet*, 2:1403.

de Kock, C. P. and Sakmann, B. (2009). Spiking in primary somatosensory cortex during natural whisking in awake head-restrained rats is cell-type specific. *Proceedings of the National Academy of Sciences of the United States of America*, 106:16446-50. doi: 10.1073/pnas.0904143106.

Descarries, L., Gisiger, V. and Steriade, M. (1997). Diffuse transmission by acetylcholine in the CNS. *Progress in Neurobiology*, 53:603-625.

Détári, L., Rasmusson, D. D. and Semba, K. (1997). Phasic relationship between the activity of basal forebrain neurons and cortical EEG in urethane-anesthetized rat. *Brain Research*, 759:112-21.

Détári, L, Rasmusson, D. D. and Semba, K. (1999). The role of basal forebrain neurons in tonic and phasic activation of the cerebral cortex. *Progress in Neurobiology*, 58:249–277.

Disney, A. A. and Aoki, C. (2008). Muscarinic acetylcholine receptors in macaque V1 are most frequently expressed by parvalbumin-immunoreactive neurons. *Journal Comparative Neurology*, 507:1748-62. doi: 10.1002/cne.21616.

Disney, A. A., Aoki, C. and Hawken, M. J. (2007). Gain modulation by nicotine in macaque V1. *Neuron*, 56:701-713.

Donoghue, J. P. and Carroll, K. L. (1987). Cholinergic modulation of sensory responses in rat primary somatic sensory cortex. *Brain Research*, 408:367–71.

Dringenberg, H. C. and Vanderwolf, C. H. (1997). Neocortical activation: modulation by multiple pathways acting on central cholinergic and serotonergic systems. *Experimental Brain Research*, 116:160–174.

Dringenberg, H. C., Hamze, B., Wilson, A., Speechley, W. and Kuo, M. C. (2007). Heterosynaptic facilitation of in vivo thalamocortical long-term potentiation in the adult rat visual cortex by acetylcholine. *Cerebral Cortex*, 17:839-48.

Duque, A., Balatoni, B., Detari, L., and Zaborszky, L. (2000). EEG correlation of the13discharge properties of identified neurons in the basal forebrain. *Journal Neurophysiology,* 84:141627-1635.

Eggermann, E. and Feldmeyer, D. (2009). Cholinergic filtering in the recurrent excitatorymicrocircuit of cortical layer 4. *Proceedings of the National Academy of Sciences of the United States of America,* 106:11753-11758.

Eggermann, E., Kremer, Y., Crochet, S. and Petersen, C. C. H. (2014). Cholinergic signals in mouse barrel cortex during active whisker sensing. *Cell Repors,* 9:1654-1660. doi: 10.1016/j.celrep.2014.11.005.

Escudero, G. and Nuñez, A. (2019). Medial Prefrontal Cortical Modulation of Whisker Thalamic Responses in Anesthetized Rats. *Neuroscience,* 406:626-636. doi: 10.1016/j.neuroscience.2019.01.059.

Fernandez de Sevilla, D. and Buño, W. (2003). Presynaptic inhibition of Schaffer collateral synapses by stimulation of hippocampal cholinergic afferent fibres. *European Journal of Neuroscience* 17:555-460.

Fibiger, H. C. (1991). Cholinergic mechanisms in learning, memory and dementia: a review of recent evidence. *Trends Neuroscience,* 14:220–223.

Fournier, G. N., Semba, K. and Rasmusson, D. D. (2004). Modality- and region-specific acetylcholine release in the rat neocortex. *Neuroscience,* 126: 257-262. doi.org/10.1016/j.neuroscience.2004.04.002.

Freund, T. F. and Meskenaite, V. (1992). gamma-Aminobutyric acid-containing basal forebrain neurons innervate inhibitory interneurons in the neocortex. *Proceedings of the National Academy of Sciences of the United States of America,* 89:738-42.

Fries, P., Reynolds, J. H., Rorie, A. E. and Desimone, R. (2001). Modulation of oscillatory neuronal synchronization by selective visual attention. *Science,* 291:1560-1563.

Gaztelu, J. M. and Buño, W. Jr. (1982). Septo-hippocampal relationships during EEG thetarhythm. *Electroencephalography and clinical neurophysiology,* 54:375-387.

Golmayo, L., Nuñez, A. and Zaborszky, L. (2003). Electrophysiological evidence for the existence of a posterior cortical-prefrontal-basal forebrain circuitry in modulating sensory responses in visual and somatosensory rat cortical areas. *Neuroscience,* 119:597-609.

Gritti, I., Mainville, L., Mancia, M., and Jones, B. E. (1997). GABAergic and other noncholinergic basal forebrain neurons, together with cholinergic neurons, project to the mesocortex and isocortex in the rat. *Journal Comparative Neurology,* 383:163-177.

Gritti, I., Manns, I. D., Mainville, L., and Jones, B. E. (2003). Parvalbumin, calbindin, or calretinin in cortically projecting and GABAergic, cholinergic, or glutamatergic basal forebrain neurons of the rat. *Journal Comparative Neurology,* 458:11-31.

Gritton, H. J., Howe, W. M., Mallory, C. S., Hetrick, V. L, Berke, J. D. and Sarter, M. (2016). Cortical cholinergic signaling controls the detection of cues. *Proceedings of the National Academy of Sciences of the United States of America,* 113:E1089-97. doi: 10.1073/pnas.1516134113.

Groenewegen, H. J. and Uylings, H. B. (2000). The prefrontal cortex and the integration of sensory, limbic and autonomic information. *Progress Brain Research,* 126:3–28.

Gruart, A., Leal-Campanario, R., López-Ramos, J.,C., Delgado-García, J. M. (2015). Functional basis of associative learning and its relationships with long-term potentiation evoked in the involved neural circuits: Lessons from studies in behaving mammals. *Neurobiology Learning and Memory,* 124:3-18. doi: 10.1016/j.nlm.2015.04.006.

Gu, Z., Alexander, G., Dudek, S. M. and Yakel, J. L. (2017). Hippocampus and entorhinal cortex recruit cholinergic and NMDA receptors separately to generate hippocampal theta oscillations. *Cell Report,* 21: 3585–3595. doi: 10.1016/j.celrep.2017.11.080.

Gulledge, A. T., Park, S. B., Kawaguchi, Y., and Stuart, G. J. (2007). Heterogeneity of phasic cholinergic signaling in neocortical neurons. *Journal Neurophysiology,* 97, 2215–2229. doi: 10.1152/jn.00493.2006.

Hampel, H., Mesulam, M. M., Claudio-Cuello, A., Farlow, M. R., Giacobini, E., Grossberg, G. T., Khachaturian, A. S., Vergallo, A., Cavedo, E., Snyder, P. J. and Khachaturian, Z. S., for the Cholinergic System Working Group. (2018). The cholinergic system in the pathophysiology and treatment of Alzheimer's disease. *Brain,* 141; 1917–1933. doi: 10.1093/brain/awy132.

Hangya, B., Borhegyi, Z., Szilágyi, N., Freund, T. F. and Varga, V. (2009). GABAergic neurons of the medial septum lead the hippocampal network during theta activity. *Journal Neuroscience,* 29:8094-102. doi: 10.1523/JNEUROSCI.5665-08.2009.

Hasselmo, M. E. (2006). The role of acetylcholine in learning and memory. *Current Opinion Neurobiology,* 16: 710–715.

Hasselmo, M. E. and Bower, J. M. (1992). Cholinergic suppression specific to intrinsicnot afferent fiber synapses in rat piriform (olfactory) cortex. *Journal Neurophysiology,* 67:1222–1229.

Hasselmo, M. E., and Sarter, M. (2011). Modes and models of forebrain cholinergic26neuromodulation of cognition. *Neuropsychopharmacology,* 36:52-73. doi: 2710.1038/npp.2010.104.

Hasselmo, M. E., Schnell, E. and Barkai, E. (1995). Dynamics of learning and recall at excitatory recurrent synapses and cholinergic modulation in rat hippocampal region CA3. *Journal Neuroscience,* 15: 5249–5262.

Herrero, J. L., Roberts, M. J., Delicato, L. S., Gieselmann, M. A., Dayan, P. and Thiele, A. (2008). Acetylcholine contributes through muscarinic receptors to attentional modulation in V1. *Nature,* 454:1110-1114.

Hoover, W. B. and Vertes, R. P. (2007). Anatomical analysis of afferent projections to the medial prefrontal cortex in the rat. *Brain Structure and Function,* 212:149-179.

Houser, C. R., Crawford, G. D., Salvaterra, P. M, and Vaughn, J. E, (1985), Immunocytochemical localization of choline acetyltransferase in rat cerebral cortex: a study of cholinergic neurons and synapses. *Journal Comparative Neurology,* 234:17-34.

Howe, W. M., Berry, A. S., Francois, J., Gilmour, G., Carp, J. M., Tricklebank, M., Lustig, C. and Sarter, M. (2013). Prefrontal cholinergic mechanisms instigating shifts from monitoring for cues to cue-guided performance: converging electrochemical and fMRI evidence from rats and humans. *Journal Neuroscience,* 33: 8742–8752.

Himmelheber, A. M., Sarter, M. and Bruno, J. P. (2000). Increases in cortical acetylcholine release during sustained attention performance in rats. *Cognitive Brain Research*, 9:313-325.

Jasper, H. H. and Tessier, J. (1971). Acetylcholine liberation from cerebral cortex during paradoxical REM sleep. *Science*, 172:601–602.

Jiang, S., Li, Y., Zhang, C., Zhao, Y., Bu, G., Xu, H. et al. (2014). M1 muscarinic acetylcholine receptor in Alzheimer's disease. *Neuroscience Bulleting*, 30: 295–307.

Jimenez-Capdeville, M. E. and Dykes, R. W. (1996). Changes in cortical acetylcholine release in the rat during day and night: differences between motor and sensory areas. *Neuroscience* 71:567-579.

Kim, J.-H., Jung, A.-H., Jeong, D., Choi, I., Kim, K., Shin, S., Kim, S. J., and Lee, S.-H. (2016). Selectivity of neuromodulatory projections from the basal forebrain and locus ceruleus to primary sensory cortices. *Journal Neuroscience* 36:5314-5327. doi: 10.1523/JNEUROSCI.4333-15.2016.

Kristt, D. A., McGowan, R. A., Martin-MacKinnon, N. and Solomon, J. (1985). Basal forebrain innervation of rodent neocortex: studies using acetylcholinesterase histochemistry, Golgi and lesion strategies. *Brain Research*, 337:19-39.

Lee, J. E., Cho, K. H., Song, S. K., Kim, H. J., Lee, H. S., Sohn, Y. H., et al. (2014). Exploratory analysis of neuropsychological and neuroanatomical correlates of progressive mild cognitive impairment in Parkinson's disease. *Journal Neurology, Neurosurgery* and *Psychiatry*, 85: 7–16.

Levey, A. I., Kitt, C. A., Simonds, W. F., Price, D. L. and Brann, M. R. (1991). Identification and localization of muscarinic acetylcholine receptor proteins in brain with subtype-specific antibodies. *Journal Neuroscience*, 11:3218-3226.

Lisman, J. E. and Jensen, O. (2013). The theta-gamma neural code. *Neuron*, 77:1002-1016.

Liu AK, Chang RC, Pearce RK, Gentleman SM. (2015). Nucleus basalis of Meynert revisited: anatomy, history and differential involvement in Alzheimer's and Parkinson's disease. *Acta Neuropathol* 129: 527–40.

Lysakowski, A., Wainer, B. H., Bruce, G. and Hersh, L. B. (1968). An atlas of the regional and laminar distribution of choline acetyltransfe-rase immunoreactivity in rat cerebral cortex. *Neuroscience,* 28:291–336.

Madoz, P. and Reinoso-Suarez, F. (1968) Influence of lesions in preoptic region on the states of sleep and wakefulness. *Proceedings XXIV International Congress Physiology and Science,* 7:276.

Manns, I. D., Alonso, A. and Jones, B. E. (2000a). Discharge properties of juxtacellularly labeled and immunohistochemically identified cholinergic basal forebrain neurons recorded in association with the electroencephalogram in anesthetized rats. *Journal Neuroscience,* 20:1505–1518.

Manns, I. D., Alonso, A. and Jones, B. E. (2000b). Discharge profiles of juxtacellularly labeled and immunohistochemically identified GABAergic basal forebrain neurons recorded in association with the electroencephalogram in anesthetized rats. *Journal Neuroscience,* 20:9252–9263.

Marrosu, F., Portas, C., Mascia, M. S., Casu, M. A., Fa, M., Giagheddu, M., Imperato, A. and Gessa, G. L. (1995). Microdialysis measurement of cortical and hippocampal acetylcholine release during sleep-wake cycle in freely moving cats. *Brain Research,* 671, 329–332.

Martin-Cortecero, J. and Nuñez, A. (2014). Tactile response adaptation to whisker stimulation in the lemniscal somatosensory pathway of rats. *Brain Research,* 1591:27-37.

Mash, D. C., Flynn, D. D. and Potter, L. T. (1985). Loss of M2 muscarine receptors in the cerebral cortex in Alzheimer's disease and experimental cholinergic denervation. *Science,* 228:1115–1117.

McCormick, D. A. (1992). Neurotransmitter actions in the thalamus and cerebral cortex and their role in neuromodulation of thalamocortical. *Progress in Neurobiology,* 39:364-370.

McCormick, D. A. and Prince, D. A. (1986). Mechanisms of action of acetylcholine in theguinea-pig cerebral cortex in vitro. *Journal Physiology,* 375:169-194.

McCormick, D. A., McGinley, M. J. and Salkoff, D. B. (2015). Brain state dependent activity in the cortex and thalamus. *Current Opinion in Neurobiology* 31:133-140. doi.org/10.1016/j.conb.2014.10.003.

McGinty, D. J. and Sterman, M. B. (1968). Sleep suppression after basal forebrain lesions in the cat. *Science,* 160:1253–1255.

Mesulam, M. M., Mufson, E. J., Wainer, B. H. and Levy, A. I. (1983). Central cholinergic pathways in the rat: an overview based on an alternative nomenclature (Ch1-Ch6). *Neuroscience*, 10:1185–1201.

Mesulam, M. M., Mash, D., Hersh, L., Bothwell, M., and Geula, C. (1992). Cholinergic innervation of the human striatum, globus pallidus, subthalamic nucleus, substantia nigra, and red nucleus. *Journal Comparative Neurology,* 323:252-268.

Metherate, R., Tremblay, N. and Dykes, R. W. (1987). Acetylcholine permits long-term enhancement of neuronal responsiveness in cat primary somatosensory cortex. *Neuroscience,* 22:75–81.

Metherate, R., Tremblay, N. and Dykes, R. W. (1988). Transient and prolonged effects of acetylcholine on responsiveness of cat somatosensory cortical neurons. *Journal Neurophysiology* 59:1253-1276.

Minces, V., Pinto, L, Danb, Y. and Chib, A. A. (2017). Cholinergic shaping of neural correlations. *Proceedings of the National Academy of Sciences of the United States of America,* 114:5725-5730; doi.org/10.1073/pnas.1621493114.

Miller, E. K. and Cohen, J. D. (2001). An integrative theory of prefrontal cortex function. *Annual Review Neuroscience*, 24:167–202.

Moruzzi, G. and Magoun, H. W. (1949). Brain stem reticular formation and activation of the EEG. *Electroencephalography Clinical Neurophysiology*, 1:455–473.

Nabavi, S., Fox, R., Proulx, C. D., Lin, J. Y., Tsien, R. Y. and Malinow, R. (2014). Engineering a memory with LTD and LTP. *Nature,* 511:348-352. doi: 10.1038/nature13294.

Nakajima, M., Schmitt, L. I. and Halassa, M. M. (2019). Prefrontal Cortex Regulates Sensory Filtering through a Basal Ganglia-to-Thalamus

Pathway. *Neuron*, 103:445-458.e10. doi: 10.1016/j.neuron.2019.05. 026.

Nieuwenhuys, R. (2009). The structural organization of the forebrain: a commentary on the papers presented at the 20th Annual Karger Workshop 'Forebrain evolution in fishes'. *Brain Behavioral and Evolution*, 74:77-85. doi: 10.1159/000229014.

Nordberg, A. and Winblad, B. (1986). Reduced number of [3H]nicotine and [3H]acetylcholine binding sites in the frontal cortex of Alzheimer brains. *Neuroscience Letters*, 72: 115–19.

Nuñez, A. (1996). Unit activity of rat basal forebrain neurons: Relationship to cortical activity. *Neuroscience*, 72:757-766.

Nuñez, A., Garcia-Austt, E. and Buño, W. Jr. (1987). Intracellular theta-rhythm generation in identified hippocampal pyramids. *Brain Research*, 416:289-300.

Nuñez, A., Garcia-Austt, E. and Buño, W. Jr. (1990). Slow intrinsic spikes recorded in vivo in rat CA1-CA3 hippocampal pyramidal neurons. *Experimental Neurology*, 109:294-299.

Nuñez, A., Dominguez, S., Buño, W. and Fernandez de Sevilla, D. (2012). Cholinergic-mediated response enhancement in barrel cortex layer V pyramidal neurons. *Journal Neurophysiology*, 108:1656-1668.

Obermayer, J., Verhoog, M. B., Luchicchi, A. and Mansvelder, H. D. (2017). Cholinergic modulation of cortical microcircuits is layer-specific: Evidence from rodent, monkey and human brain. *Frontiers in Neural Circuits*, 11:100.

Oldford, E. and Castro-Alamancos, M. A. (2003). Input-specific effects of acetylcholine on sensory and intracortical evoked responses in the "barrel cortex" in vivo. *Neuroscience*, 117:769-778.

Parikh, V. and Sarter, M. (2008). Cholinergic mediation of attention: contributions of phasic and tonic increases in prefrontal cholinergic activity. *Annual New York Academia Science*, 1129: 225–235.

Parikh, V., Pomerleau, F., Huettl, P., Gerhardt, G. A., Sarter, M. and Bruno, J. P. (2004). Rapid assessment of in vivo cholinergic transmission by amperometric detection of changes in extracellular choline levels. *European Journal Neuroscience*, 20:1545–1554.

Parikh, V., Kozak, R., Martinez, V. and Sarter, M. (2007). Prefrontal acetylcholine releasecontrols cue detection on multiple timescales. *Neuron*, 56: 141–154.

Pinto, L., Goard, M. J., Estandian, D., Xu, M., Kwan, A. C., Lee, S. H., Harrison, T. C., Feng, G. and Dan, Y. (2013). Fast modulation of visual perception by basal3forebrain cholinergic neurons. *Nature Neuroscience*, 16:1857-1863. doi: 10.1038/nn.3552.

Qian, J. and Saggau, P. (1997). Presynaptic inhibition of synaptic transmission in the rat hippocampus by activation of muscarinic receptors: involvement of presynaptic calcium influx. *British Journal of Pharmacology*, 122:511-519.

Radnikow, G. and Feldmeyer, D. (2018). Layer- and Cell Type-Specific Modulation of Excitatory Neuronal Activity in the Neocortex. *Frontiers in Neuroanatomy*, 30:12:1. doi: 10.3389/fnana.2018.00001.

Rasmusson, D. D. (2000). The role of acetylcholine in cortical synaptic plasticity. *Behavioral Brain Research*, 115:205–218.

Rasmusson, D. D., Clow, K., and Szerb, J. C. (1992). Frequency-dependent increase in cortical acetylcholine release evoked by stimulation of the nucleus basalis magnocellularis in the rat. *Brain Research*, 594:150–154.

Ray, N. J., Bradburn, S., Murgatroyd, C., Toseeb, U., Mir, P., Kountouriotis, G. L., Teipel, S. J. and Grothe, M. J. (2018). In vivo cholinergic basal forebrain atrophy predicts cognitive decline in de novo Parkinson's disease. *Brain*, 141; 165–176.

Rodriguez, R., Kallenbach, U., Singer, W. and Munk, M. H. (2004). Short- and long-term effects of cholinergic modulation on gamma oscillations and response synchronization in the visual cortex. *Journal Neuroscience*, 24:10369-10378.

Sarter, M. and Bruno, J. P. (1997). Cognitive functions of corticalacetylcholine: toward a unifying hypothesis. *Brain Research Review*, 23:28–46.

Sarter, M., Nelsonm C. L. and Bruno, J. P. (2005). Cortical cholinergic transmission and cortical information processing in schizophrenia. *Schizophrenia Bulletin*, 31:117–138.

Sarter, M., Parikh, V. and Howe, W. M. (2009). Phasic acetylcholine release and the volume transmission hypothesis: time to move on. *Nature Review Neuroscience,* 10:383–390.

Sarter, M., Lustig, C., Howe, W. M., Gritton, H. and Berry, A. S. (2014). Deterministic functions of cortical acetylcholine. *European Journal Neuroscience,* 39:1912–1920.

Saunders, A., Granger, A. J. and Sabatini, B. L. (2015). Corelease of acetylcholine and GABA from cholinergic forebrain neurons. *Elife,* 4:e06412. doi: 10.7554/eLife.06412.

Scheef, L., Grothe, M. J., Koppara, A., Daamen, M., Boecker, H., Biersack, H., Schild, H. H., Wagner, M., Teipel, S. and Jessen, F. (2019). Subregional volume reduction of the cholinergic forebrain in subjective cognitive decline (SCD). *Neuroimage Clinical,* 21:101-112.

Schroder, H., Giacobini, E., Struble, R. G., Zilles, K.and Maelicke, A. (1991) Nicotinic cholinoceptive neurons of the frontal cortex are reduced in Alzheimer's disease. *Neurobiology Aging,* 12: 259–262.

Shute, C. C. and Lewis, P. R. (1975). Cholinergic and pathways. *Pharmacology and Therapheutics B,* 1:79-87.

Sillito, A. M. and Kemp, J. A. (1983). Cholinergic modulation of the functional organization of the cat visual cortex. *Brain Research,* 289:143–155.

Singer, W. (1999). Neuronal synchrony: A versatile code for the definitions of relations? *Neuron,* 24:49–65.

Soma, S., Shimegi, S., Suematsu, N., Tamura, H., Sato, H. (2013). Modulation-specific and laminar-dependent effects of acetylcholine on visual responses in the rat primary visual cortex. *PLoS One,* 8:e68430. doi: 10.1371/journal.pone.0068430.

Spencer, J. P., Middleton, L. J. and Davies, C. H. (2010). Investigation into the efficacy of the acetylcholinesterase inhibitor, donepezil, and novel procognitive agents to induce gamma oscillations in rat hippocampal slices. *Neuropharmacology,* 59:437-443.

Stoiljkovica, M., Kelleya, C., Nagya, D., Leventhalb, L. and Hajos, M. (2016). Selective activation of α7 nicotinic acetylcholine receptors augments hippocampal oscillations. *Neuropharmacology*, 110: 102e108. doi.org/10.1016/j.neuropharm.2016.07.010.

Szymusiak, R. and McGinty, D. (1986). Sleep suppression following kainic acid-induced lesions of the basalforebrain. *Experimental Neurology*, 94:598–614.

Takács, V. T., Cserép, C., Schlingloff, D., Pósfai, B., Szőnyi, A., Sos, K. E., Környei, Z., Dénes, Á., Gulyás, A. I., Freund, T. F. and Nyiri, G. (2018). Co-transmission of acetylcholine and GABA regulates hippocampal states. *Nature Communications*, 9:2848. doi: 10.1038/s41467-018-05136-1.

Unal, C. T., Golowasch, J. P. and Zaborszky, L. (2012). Adult mouse basal forebrain harbors two distinct cholinergic populations defined by their electrophysiology. *Frontiers Behavioral Neuroscience*, 6:21. doi: 10.3389/fnbeh.2012.00021.

Verhoog, M. B., Obermayer, J., Kortleven, C. A., Wilbers, R., Wester, J., Baayen, J. C., De Kock, C. P. J., Meredith, R. M. and Mansvelder, H. D. (2016). Layer-specific cholinergic control of human mouse cortical synaptic plasticity. *Natural Communication*, 7:12826. doi: 10,1038/ncomms12826.

Vertes, R. P. (2005). Hippocampal theta rhythm: a tag for short-term memory. *Hippocampus*, 15:923-935.

Vertes, R. P. and Kocsis, B. (1997). Brainstem-diencephalo-septohippocampal systems controlling the theta rhythm of the hippocampus. *Neuroscience*, 81:893–926.

Whitehouse, P. J., Price, D. L., Struble, R. G., Clark, A. W., Coyle, J. T. and Delon, M. R. (1982). Alzheimer's disease and senile dementia: loss of neurons in the basal forebrain. *Science*, 1982; 215: 1237–8.

Wimmer, R. D., Schmitt, L., Nakajima, M., Deisseroth, K. and Halassa, M. M. (2015). Thalamic control of Sensory selection in divided attention. *Nature*, 526:705-709.

Womelsdorf, T., Fries, P., Mitra, P. P. and Desimone, R. (2006). Gamma-band synchronization in visual cortex predicts speed of change detection. *Nature*, 439:733-736.

Zaborszky, L. (2002). The modular organization of brain systems. Basal forebrain: the last frontier. *Progress Brain Research*, 136:359-370.

Zaborszky, L., and Duque, A. (2000). Local synaptic connections of basal forebrain neurons. *Behavioral Brain Research*, 115:143–158. doi: 10.1016/s0166-4328(00)00255-2.

Zaborszky, L., Gaykema, R. P., Swanson, D. J., and Cullinan, W. E. (1997). Cortical input to the basal forebrain. *Neuroscience* 79:1051–1078. doi: 10.1016/s0306-4522(97)00049-3.

Zaborszky, L., Buhlm, D. L., Pobalashinghamm, S., Bjaaliem, J. G., and Nadasym, Z. (2005). Three-dimensional chemoarchitecture of the basalforebrain: spatially specific association of cholinergic and calcium binding-protein-containing neurons. *Neuroscience*, 136:697-713.

Zaborszky, L., Csordas, A., Mosca, K., Kim, J., Gielow, M. R., Vadasz, C., et al. (2015). Neurons in the basal forebrain project to the cortex in a complextopographic organization that reflects corticocortical connectivity patterns: anexperimental study based on retrograde tracing and 3D Reconstruction. *Cerebral Cortex,* 25:118–137. doi: 10.1093/cercor/bht210.

Zhang, H., Lin, S. C., Nicolelis, M. A. (2010). Spatiotemporal coupling between hippocampal acetylcholine release and theta oscillations in vivo. *Journal Neuroscience*, 30:13431-13440.

BIOGRAPHICAL SKETCH

Angel Nuñez Molina, PhD

Affiliation: Full Professor (Cellular Biology)

Professional address: Dept. Anatomia, Histologia y Neurociencia Fac. Medicina Univ. Autonoma de Madrid

Education:
B.S. Faculty of Biological Sciences, Universidad Autonoma de Madrid.
Ph.D. in Biology (Honors), Universidad Autonoma de Madrid

Business Address: Fac. Medicina. Univ. Autónoma de Madrid Arzobispo Morcillo 4, 28029 Madrid. Spain

Research and Professional Experience:

Research Experience:
1987-present, Research collaborator and Principal Investigator (since 2002) in the following Grant fields: Somatosensory information processing and neurophatic pain. Slow waves generation in the EEG. Basal forebrain regulation of the cortical activity.

Teaching Experience:
- 1987-present: Participation in undergraduate programs of Cellular Biology and Neuroscience for medical students.
- 1982-present: Participation in graduate courses within Ph.D. Programs of Neuroscience in Spain (Universidad Autonoma de Madrid, Universidad Castilla-León).
- 2000-present Direction of over twenty graduate courses within Doctoral Programs of Neuroscience.
- 2002-present: Direction of eleven Doctoral Theses, all of which received the highest rating.

Professional Appointments: Neurophysiology of the somatosensory system and sleep-wakefulness regulation. At present my main interest is the somatosensory information processing in the cortex and its modulation by acetyl choline and the grow factor IGF-I.
Ih index: 35 Google Scholar (2019-08-08).

Honors:
- 2004 Short-term fellowship in the Laboratoire de Neurophysiologie, Faculté de Medicine, Université Laval, Quebec, Canada.
- 1990-1992. Postdoctoral fellowship in the Laboratoire de Neurophysiologie, Faculté de Medicine, Université Laval, Quebec, Canada.
- 1986 and 1987. Research fellowship from "Fondo de InvestigacionesSanitarias (FISS)".
- 1983-1985. Research fellowship from "Comisión de Investigación delSíndrome Tóxico".

Publications from the Last 3 Years:
1. Diez-Garcia, A., Barros-Zulaica N., NúñezA., Buño W. and Fernandez de Sevilla D. (2017). Bidirectional Hebbian plasticity induced by low-frequency stimulation in basal dendrites of rat barrel cortex layer 5 pyramidal neurons. *Frontiers CellularNeuroscience.* 11:8. doi.org/10.3389/fncel.2017.00008.
2. Stein, A. M., Munive, V., Fernandez, A. M., Núñez A. and Torres-Aleman I. (2017). Acute exercise does not modify brain activity and memory performance in APP/PS1 mice. *PLoS ONE* 12(5): e0178247. doi.org/10.1371/journal.pone.0178247.
3. Robledinos-Antón, N., Rojo, A. I., Ferreiro, E., Núñez, A., Krause, K. H., Jaquet, V., Cuadrado, A. (2017).Transcription factor NRF2 controls the fate of neural stem cells in the subgranular zone of the hippocampus. *Redox Biology* 13:393-401. doi: 10.1016/j.redox.2017.06.010.
4. Rojo, A. I., Pajares, M., Rada, P., Núñez, A., Nevado-Holgado, A., Killik, R., VanLeuven, F., Ribe, E., Lovestone, S., Yamamoto, M., Cuadrado A. (2017). NRF2deficiency replicates transcriptomic changes in Alzheimer's patients and worsens APP and TAU pathology. *Redox Biology* 13:444-451.doi: 10.1016/j.redox.2017.07.006.
5. Casas-Torremocha, D., Clasca, F. and Núñez, A. (2017). Plasticity of tactile responses mediated by the posterior medial thalamic

nucleus in rat motor and somatosensory cortices. *Frontiers Neural Circuits* 11:69.doi: 10.3389/fncir.2017.00069.
6. Reyes-Marín, K. E. and Núñez, A. (2017). Seizure susceptibility in a transgenic animal model of Alzheimer's disease and relationship with amyloid β plaques. *Brain Research* 1677: 93-100. doi: 10.1016/j.brainres.2017.09.026.
7. Chaves-Coira, I., Rodrigo-Angulo, M. L. and Núñez, A. (2018). Bilateral pathways from the basal forebrain to sensory cortices may contribute to synchronous sensory processing. *Frontiers in Neuroanatomy* 12:5. doi: 10.3389/fnana.2018.00005.
8. Chaves-Coira, I., Martín-Cortecero, J., Núñez, A. and Rodrigo-Angulo, M. L. (2018). Different basal forebrain nuclei display distinct projecting pathways and functional circuits to sensory primary and prefrontal cortices in the rat. *Frontiers in Neuroaanatomy* 12:69. doi: 10.3389/fnana.2018.00069.
9. Escudero, G. and Núñez, A. (2019). Medial prefrontal cortical modulation of whisker thalamic responses in anesthetized rats. *Neuroscience* 4522:30086-7. doi.org/10.1016/j.neuroscience.2019.01.0599.
10. Pereda-Pérez, I., Valencia, A., Baliyan, S., Núñez, A., Sanz-García, A., Zamora-Crespo, B., Rodríguez-Fernández, R., Esteban, J. A. and Venero, C. (2019). Systemic administration of a fibroblast growth factor receptor 1 agonist rescues the cognitive deficit in aged socially isolated rats. *Neurobiology of Aging* 78: 155-165. doi.org/10.1016/j.neurobiolaging.2019.02.011.
11. Casas-Torremocha, D., Prorrero, C., Rodriguez-Moreno, J., García-Amado, M., Lübke, J. H. R., Núñez, A. and Clasca, F. (2019). Posterior thalamic nucleus axons have different terminal structures and functional impact in the vibrissal motor and somatosensory cortices. *Brain Structure and Function.* doi.org/10.1007/s00429-019-01862-4.
12. Barros-Zulaica, N., Villa, A., and Núñez, A. (2019). *Response adaptation in barrel cortical neurons facilitates stimulus detection during rhythmic whisker stimulation in anesthetized mice.*

eNeuro6(2) e0471-18.2019 1–15.doi.org/10.1523/ENEURO.0471-18.2019.

13. Zegarra-Valdivia, J. A., Santi, A., Fernandez de Sevilla, M. E., Núñez, A. and Torres-Aleman, I. (2019). Serum Insulin-Like Growth Factor I Deficiency Associates to Alzheimer's Disease Co-Morbidities. *Journal of Alzheimer's Disease* 69:979-987. doi: 10.3233/JAD-190241.

In: The Forebrain
Editor: Morten F. Thorsen
ISBN: 978-1-53618-407-5
© 2020 Nova Science Publishers, Inc.

Chapter 3

THE ROLE OF REELIN IN CORTEX-DEPENDENT MOTOR FUNCTION

Mariko Nishibe[1,2,*] *and Yu Katsuyama*[2]

[1]Graduate School of Dentistry Osaka University, Osaka, Japan
[2]Department of Anatomy Shiga University of Medical Science, Shiga, Japan

ABSTRACT

Reelin has been identified as a gene responsible for the abnormal phenotype observed in *reeler*, the classic mouse mutant strain that exhibits some functional impairments reminiscent of these found in neurological disorders. Reelin is an extracellular glycoprotein which regulates and comprises intracellular biochemical signaling through binding to its cell membrane receptors. Studies undertaken in the last two decades, using human brain specimens and human genome data, have suggested the involvement of *REELIN* in various brain pathologies. Early reports on abnormalities in *reeler* described ataxia attributed to cerebellar dysfunction and disruption of laminar structure in the cerebral cortex. More recent studies on *reeler*, on the other hand, have described detailed abnormalities

---

* Corresponding Author: Mariko Nishibe, Address: 1-8 Yamada-oka, Suita, Osaka 565-0871, Japan, Email: mnishibe@dent.osaka-u.ac.jp.

of the cortex at the level of synaptic functions, neuronal plasticity and dendritic spine formation. In this review, we discuss the putative involvement of Reelin signal in motor-related impairments observed in neurological diseases, including lissencephaly, psychiatric disorders and brain injuries.

Keywords: cerebral cortex, Reelin, lissencephaly, psychiatric diseases, brain injury, motor function

INTRODUCTION

Reelin is required for the migration of differentiating cortical neurons to arrive in an appropriate layer from the ventricular zone where the neurons are born during brain development, forming an inside-out neuronal alignment along the cortical radial axis in rodents (Caviness and Rakic 1978). Thus, when the genetic locus (*RELN*) is affected, through spontaneous autosomal recessive mutation in mice, laminar structures, of the cerebral cortex, cerebellar cortex, olfactory bulb, and hippocampus become disorganized. Not only the laminar structures but the mutation also affects non-laminar structures, including inferior olivary complex, facial nucleus, and the motor trigeminal nucleus (Terashima 1995a, 1995b), the mechanism of which remains unknown. Further, Reelin continues to be secreted in the adult brain; and hence, a synaptic modulatory role has been demonstrated in mice (Chen et al. 2005; Qiu et al. 2006). In some brain disorders including lissencephaly and developmental psychiatric disorders, deterioration in Reelin level has been associated at the genetic or at the protein level. Diverse clinical observations of patients with these brain disorders are at least partially accounted for by improper migration of neurons, or impaired synaptic transmission and plasticity caused by the lack of proper function of Reelin. Thus, empirical examination of the phenotype *Reelin*-deficient mutant (hereafter *reeler*) mice display offers inferences about various brain functions. Aiming to elucidate the neuronal substrate within the cortical network, we have, in particular, carefully examined the motor impairments displayed by homozygous and heterozygous *reeler* mice. In the present review, after briefly introducing the glycoprotein Reelin (I), we discuss the role of Reelin in the context of motor-related behaviors (II-IV).

I. REELIN SIGNALING PATHWAY AND *REELER* MOUSE

Reeler is a spontaneously occurred mouse mutant strain, described by Falconer in 1951 (Falconer 1951). *Reelin*, the gene responsible for the *reeler* mutant phenotype was identified at chromosome 7q22 (D'Arcangelo et al. 1995). Murine *RELN* gene consists of 65 exons, spanning approximately 450 kb (Royaux et al. 1997). The Reelin protein is a large extracellular matrix glycoprotein, and its serine protease activity was reported (Quattrocchi et al. 2002). N-terminal of the Reelin protein contains a signaling peptide, followed by a region of which the amino acid sequence is similar to F-spondin (so called F-spondin domain). After un-conserved region, 8 repeats of 300-350 amino acids containing an epidermal growth factor motif at their center follows. Each of this repeat can be divided into two sub-repeats: A (the BNR/Asp-box repeat) and B (the EGF-like domain). The C-terminal part of the protein contains a short C-terminal region (CTR). The amino acids sequence of CTR is almost identical in all investigated mammals, suggesting the importance of CTR in the protein function. Nakano et al. (2007) examined mutated Reelin protein from which the CTR was removed and found that while this truncated Reelin is still secreted from cells, the efficiency to phosphorylate Dab1 the downstream adaptor protein was much lower, suggesting that CTR is required for activating downstream events of the signaling pathway (Nakano et al. 2007). Reelin is cleaved *in vivo* at two sites, producing three fragments (Lambert de Rouvroit et al. 1999). The central fragment containing Reelin repeats 3–6 can bind to the Reelin specific receptors, and overexpressing the central fragment can recapitulate the functions of the full-length protein during cortical plate development.

When both alleles of the genetic locus *REELIN* (*RELN*) are affected, laminar structures, including the cerebral cortex, cerebellar cortex, olfactory bulb, and hippocampus, become disorganized, as observed in homozygous *reeler* mice. The mouse mutants, *yotari* and *scrambler* have abnormal brain morphology indistinguishable from that of homozygous *reeler* (Sheldon et al. 1997; Yamamoto et al. 2009; Ware et al. 1997). *Dab1*, a gene related to the Drosophila gene *'disabled'* (*dab*), was identified as the affected gene in *yotari* and *scrambler* mutants (Ware et al. 1997; Sheldon et al. 1997). Moreover, in the homozygous *reeler*, neurons in the developing cortex accumulate approximately 10-fold more Dab1 protein than the neurons cultured from their wildtype (WT) counterpart, suggesting degradation of Dab1 is dependent on the Reelin signal activation (Feng et al. 2007). Thus, the signaling transduction to intracellular molecules requires the phosphorylation of Dab1, whereas degradation of Dab1 should attenuate activation of the signaling. Although the stoichiometry of these outcomes of Reelin signal activation on Dab1 protein is not yet clear, this negative feedback may be important for precise determination of the duration of Reelin signal activation.

Upon binding of Reelin to its receptors—Low density lipoprotein receptor-related protein 8 (LRP8, also known as apolipoprotein E receptor 2, ApoER2) and Very low density lipoprotein receptor (VLDLR)— Dab1 protein becomes phosphorylated by Src family kinases. Knockout mice lacking both of these receptor genes *Lrp8* and *Vldlr* precisely mimic the phenotype of *reeler* and *yotari/ scrambler* mice, including the disorganized cortical layers and absence of cerebellar foliation (Trommsdorff et al. 1999). Furthermore, the mutant mice in which Reelin-induced phosphorylation sites of Dab1 were silenced showed brain abnormalities that are similar to the Dab1 null mutants (Howell et al. 2000). These experimental results have now established the Reelin-Dab1 signaling pathway. Many other molecules have been reported as downstream factors affecting this signaling pathway, resulting in cell migration, neurite development, and synaptic function (Figure 1).

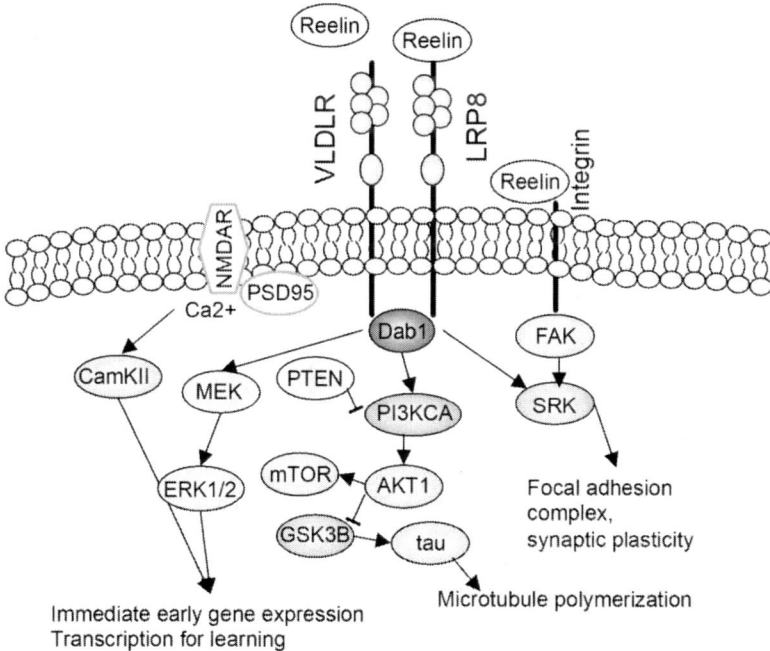

Figure 1. Reelin signaling pathway.

II. MOTOR FUNCTION IN LISSENCEPHALY

Reelin plays multiple roles in the cerebral cortex. The primary and most well-known role of Reelin is in regulation of neural migration along the radial axis of the developing cerebral cortex (Caviness and Rakic 1978). Reelin is expressed in the neurons called Cajal-Retzius cells, the neurons that differentiate earliest during the cortical development. Despite its prominent appearance, the function of Cajal-Retzius cells was not known before the discovery of the Reelin expression. Because *reeler* mice exhibited disrupted morphology in the cerebral cortex, the Cajal-Retzius cell was found to contribute to the formation of the six-layered structure through the secretion of Reelin protein.

Clinical heterogeneous manifestations of lissencephaly in terms of severity and affected structure suggest that causal mutation cannot be a single gene; nor can clear genotype and phenotype correlation be currently identified (Tan, Chong, and Mankad 2018; Devisme et al. 2012). For example, lissencephaly related genes include *LIS1, CX, ACTB, ACTG1, ARX, CK5, CRADD, DYNC1H1, KIF2A, KIF5C, NDE1/NDEL1, TUBA1A, TUBA8, TUUBB, TUBB2B, TUBB3, TUBG1, RELN and VLDLR*; and these genes are expressed in various percentages of patients with lissncephaly. As such, the human reelin (*RLEN*) is one of the genes found in some patients with lissencephaly (Hong et al. 2000; Guerrini and Dobyns 2014). The *RELN* mutation on chromosome7p22 disrupts splicing of *RELN* cDNA that reduces the reelin protein to an undetectable level (D'Arcangelo et al. 1995). The null allele carriers manifest lissencephaly with severe abnormalities of the cerebellum, hippocampus and brainstem. Lissencephaly is a developmental disorder in which the long-range neural migration in the developing cerebral cortex is interrupted, resulting in a thickened cerebral cortex with smooth and simplified folding. The null mutation in one of the two cellular surface receptors for reelin, *VLDLR*, has also been identified to manifest the cortical malformation with cerebellar hypoplasia in humans (Boycott et al. 2005; Kolb et al. 2010). Children with this malformation of cortical development require medical attention for issues ranging from early feeding problems to gross developmental delay, exhibiting continued growth hindrance (Guerrini and Dobyns 2014; Ross, Swanson, and Dobyns 2001). Because of severe developmental psychomotor consequence of the null mutation causing lissencephaly with cerebellar hypoplasia, extensive reports on their motor control have not been available other than the general descriptions on hypotonia, severe ataxia, and seizures in patients identified with *RELN* mutation.

Similarly, the genetic consequence of murine *RELN* mutation on chromosome 5 exerts disruption in the cortical layer development (D'Arcangelo et al. 1995). In pups, by PND 14, a substantial developmental deterrence is noticed, compared to the heterozygous and WT pups. The homozygous reeler pups require feeding attention and in some cases require isolation from other pups to prevent weaning competition. In an early report by Myers (1970), the mutants were identified among other littermates through observation of motor defects (Myers 1970). In spite of the successive decline in the survival rate of mutants, "healthy" *reeler* mice can survive after maturation. The mutant males can produce litters with normal females, and so can female mutants with normal males. The mutant mice of both sexes can care for their pups, such as suckling, grooming, and retrieving of pups, and can build nests. Thus, the motor impairments induced by Reelin-deficiency do not bar these mice from reproduction. Because the mutant mice are observed constantly smaller than normal mice after postnatal day 12, it is possible they suffer from gastro-intestinal problems. The reduction in food intake may be due to an indirect effect of the motor syndrome. Myers notes that, even as almost all homozygous *reeler* mice can survive until 20 days after birth, very few *reeler* could survive after 200 days after birth. Since the first stock of *reeler* mice began on C47BL/6J strain, the homozygous mutation was associated with low viability at the weaning, which to some extent was considered dependent on the genetic background of inbred strain (Caviness, So, and Sidman 1972). Our recent experience suggests homozygous *reeler* mice (on B6C3Fe strain) rarely die after maturity when weaning competition is removed with adequately provided feeding. The broad lifespan of homozygous *reeler* suggests the lethality may not only be based on the genetic background but also on the quantity of food intake, motor syndrome and the complications frequently dependent on how mice are kept in the husbandry. Further, cognitive function of *reeler* mice has not been precisely described which might contribute to the cause of earlier death in early postnatal days compared to the WT and heterozygous mice.

To further evaluate the importance of the cortical laminar structure in behavior, we constructed the dorsal forebrain specific knockout mouse strain of *Dab1* gene (*Dab1$^{floxed/floxed}$; Emx1-Cre*) (Imai et al. 2017). We chose *Dab1* to disrupt Reelin signal because Dab1 is an intracellular protein; and hence, specific targeting of the signaling pathway in neurons of the cerebral cortex was possible. Unlike homozygous *reeler* mice, death around weaning time of the cerebral cortex-specific *Dab1* knockout (*Dab1* cKO) is not more frequent than that in control mice. Perhaps, if there was any cognitive impairment accounted for that relates to early postnatal death, it does not alone induce increased mortality. As expected, *Dab1* cKO mice exhibit no gait disturbances from normal cerebellum development. Furthermore, *Dab1* cKO mice show normal grip strength, as well as normal performance on the rotarod and balance beam tests, suggesting that basal motor movements are normal even with disrupted cortical laminar when the cerebellum is intact (Imai et al. 2017).

Although basal motor movements were normal, we found that on the basis of the disrupted cortical layer formation, the adult homozygous *reeler* mice and homozygous *Dab1* cKO mice displayed impairments in the forepaw fine motor control (Nishibe et al., 2018). The dexterous functional impairment was detected in handling of a small chow pellet, an uncracked sunflower seed, and a 1.5mm length uncooked pasta. The reproduced behavioral results of homozygous *reeler* mice in homozygous *Dab1* cKO mice indicate that the impairment in the hand manipulation tasks is specific to the cortical abnormality. Neural projections were reported normal from the cortex to the target muscle in the homozygous mutants (Terashima 1995b). In agreement, our assessment showed the cortical motor representation in the homozygous *reeler* mice was preserved, derived by intracortical microstimulation (ICMS) (Nishibe et al., 2018). In contrast, we found that the cortical stimulation threshold to evoke joint movements was two-three times higher in the homozygous *reeler* and in homozygous *Dab1* cKO mice than in their respective controls. Cortical projection neurons are found not only in the large pyramidal cell layer but also in other cortical layers in the mutants (Terashima et al. 1992). The burst-generating cells typically found in the WT deeper layers are distributed in the superficial half of the *reeler* sensorimotor cortex (Silva, Gutnick, and Connors 1991). Thus, the cortical malformation may be explained by irregular neuronal alignment, therefore lowering the efficiency of activating cortical pyramidal neurons. It is possible a similar phenomenon occurs clinically in individuals with cortical malformation who are described to exhibit widely spaced gyri and thickened gray matter (Patel and Barkovich 2002; Zaki et al. 2007), resulting in less efficient activation of motor movements attributed to the cerebral cortex.

Studies on the heterogeneous group of human cortical malformation report the possibility of neuromuscular connectivity impairment. Often hypotonia is described as a clinical manifestation (Hourihane et al. 1993; Ross, Swanson, and Dobyns 2001). Our study using the homozygous *reeler* mice showed that in the forelimb muscle the genetic mutation did not cause any changes in the connectivity of neuromuscular junction, in the level of choline acetyltransferase, or in the muscle weight (Nishibe et al., 2018). The clinical evaluation of patients with cerebellar ataxia *with normal cortical function* testing for hypotonia or hyporeflexia showed normal muscle tone and normal tendon reflexes (Diener et al. 1992), unless the acute lesion was the cause of cerebellar ataxia. It may be that other lissencephaly-related genes may cause muscle tone impairment; or hypotonia may arise from the secondary effect of motor disorders, for example, due to disuse of the limbs or spasticity.

III. MOTOR FUNCTION IN NEURODEVELOPMENTAL PSYCHIATRIC DISORDER

Neurodevelopmental disorders including schizophrenia, autism spectrum syndrome, attention deficits hyperactive disorder (ADHD), and bipolar disorder impose on patients not only psychiatric symptoms but often also impose difficulties in motor coordination, specified as one of eight dimensions of psychopathology, in the diagnostic and statistical manual (DSM-5). Heterozygous *reeler* mutants (HRM) have been suggested to be an animal model of neurodevelopmental psychiatric disorders. The first report of clinical cases of possible involvement of Reelin is in patients with schizophrenia. *RELN* mRNA and its protein product were reduced to 50% in the cerebral cortex, hippocampus and cerebellum (Costa et al. 2002). It is noteworthy, however, that individuals with *RLEN* null alleles (who exhibit lissencephaly with hypoplasia) or with heterozygous *RELN* mutation (as in parents and unaffected siblings of the total null alleles carriers) were described not to exhibit schizophrenic-like symptoms, but null allele mutation carriers showed severe delay in cognitive development with little or no language (Hong et al. 2000). Epigenetic mechanism in downregulation of *RELN* has been postulated, but remains controversial (Mill et al. 2008; Tochigi et al. 2008). Other studies reported that a reduction in fragile X mental retardation protein (FMRP) has been identified in several psychiatric schizophrenia, bipolar, and autism that disrupts reelin translation, thereby reducing the level of protein expression (Fatemi and Folsom 2011).

It could also be a secondary effect associated with the reduction in reelin expression causing vulnerability to clinical psychosis and psychomotor dysfunction. For example, in addition to a regional reduction in the level of reelin, a reduction in the level of glutamic acid decarboxylase (GAD) 67 and turnover rate from glutamate to GABA have been documented in the prefrontal cortex of schizophrenia (Impagnatiello et al. 1998; Guidotti et al. 2000; Fatemi et al. 2005). The GABAergic dysfunction is also noted in the motor control related cortical areas, in patients with schizophrenia, (Hashimoto et al. 2008; Guidotti et al. 2000; Northoff, Steinke, et al. 1999). Though genetically affected by the *reeler* genetic locus, the heterozygous *reeler* mouse (HRM) shows reductions in GAD-67 expression and in parvalbumin-positive GABAergic interneurons (Ammassari-Teule et al. 2009). Also, GABA alteration in HRM has also been described in the prefrontal cortex (Liu et al. 2001) and in hippocampus (Nullmeier et al. 2011).

Similarly, some cognitive behavioral impairment commonly observed in individuals with psychiatric disorders (Perry, Geyer, and Braff 1999; McAlonan et al. 2002; Turetsky et al. 2007) has been found to occur in the HRM. The behavioral deficits were detected in HRM on radial maze (Carboni et al. 2004), prepulse inhibition tests (PPI) (Teixeira et al. 2011; Barr et al. 2008; Tueting et al. 1999), executive function (Krueger et al. 2006; Brigman et al. 2006), contextual fear associative learning test (Qiu et al. 2006), recognition memory task (Kutiyanawalla et al. 2012), and on olfactory-conditioned learning (Weeber et al. 2002; Rogers et al. 2011; Qiu et al. 2006; Trotter et al. 2013; Beffert et al. 2005; Ognibene et al. 2007).

The sensorimotor deficits, on the other hand, were not reported until recently in HRM. Our recent study provided insight into the possible role of Reelin-Dab1 signaling in motor skill learning that requires cortical synaptic plasticity. We observed motor learning deficits in the consecutive 10-day reach-to-grasp training in the HRM and heterozygous *Dab1* cKO mice (Nishibe et al., submitted). In both mutant genotypes, though Reelin haploinsufficiency did not prevent the motor learning ability completely, the degree of learning was lower than respective control mice. In stabilized patients with schizophrenia, the dexterous motor skills tested (for example, finger force tracking or sequential finger tapping) are impaired (Teremetz et al. 2017). A number of other studies on patients with schizophrenia also reported the motor dexterous impairments (Teremetz et al. 2014; Midorikawa et al. 2008; Gschwandtner et al. 2006; Jogems-Kosterman et al. 2006; Kern et al. 1998; Morrens, Docx, and Walther 2014; Schwartz et al. 1990). Hence, our results support the notion that the HRM could represent some behavioral deficits of a form of mental disorder. It has to be noted, however, that throughout the study, HRM did not exhibit catatonia, often described to accompany the severe forms of schizophrenia. The HRM can also perform equally well in some of the different tasks previously demonstrated as implicated with schizophrenia, including cognition and fear memory (Krueger et al. 2006; Salinger, Ladrow, and Wheeler 2003). Hence, modeling mental disorder conditions by one genetic manipulation alone imposes some limitation. Nevertheless, using the HRM as a model of neurodevelopmental disorders helps us understand the phenotype-relevant molecular mechanism.

The functional MRI in a number of studies has elucidated the cortical motor map of patients with schizophrenia, indicating comparatively reduced activity (Scheuerecker et al. 2009; Schroder et al. 1999; Schroder et al. 1995). More specifically, disturbances in the initiation of movements in the finger tapping as studied in akinetic patients were associated with primary motor area (M1) and supplementary motor area (SMA) hypofunction (Northoff, Braus, et al. 1999; Payoux et al. 2004). Furthermore, the altered functional level of the premotor areas was linked to motor skillful learning in schizophrenia (Exner et al. 2006). Our study was in agreement with the notion that motor cortical adaptation to the skill demand is altered in the Reelin haploinsufficient model. To illustrate, detailed cortical map representations were derived by ICMS in the HRM and the heterozygous *Dab1*cKO mice, with and without the reach-to-grasp training; and we found the extent of the cortical reorganization was smaller in the HRM and the heterozygous *Dab1*cKO mice, than in the respective control groups, implicating the role of Reelin in synaptic plasticity within the cortical network.

IV. MOTOR CONTROL IN BRAIN INJURY

A decrease in quality of life due to dexterous impairments can also, and largely, be found in people with brain injury, often related to trauma or to vascular-cause, from the pediatric population (Mirkowski et al. 2019) to the elderly population. Reelin involvement is suggested not only during development and adulthood, but also after brain injuries.

Study results indicate Reelin protein works for advantage in recovery against acute brain injuries. When a brain injury was induced, the lesion was larger in the homozygous *reeler* mice (Gowert et al. 2017). Because Reelin regulates the migration of progenitor cells from the subventricular zone in intact adult mice, it is suggested that Reelin is upregulated at the site of induced-injury in mice following brain lesions (Courtes et al. 2011; Massalini et al. 2009). Subsequently, when reelin was overexpressed, the lesion volume was smaller (Won et al. 2006; Courtes et al. 2011). Reducing microRNA200c, targeting *reelin*, inversely upregulating reelin before induction of transient MCAO (middle cerebral artery occlusion) were reported to ameliorate the neurological deficit and reduce the lesion volume 24 hours post transient MCAO (Stary et al. 2015).

During the chronic stage of a brain injury, patients often undergo motor rehabilitative training. In the current literatures, with or without the aid of rehabilitative training, motor recovery is possible because of brain plasticity (Hallett 2001; Jones et al. 2013; Nudo, Plautz, and Frost 2001). Our previous studies showed that extensive and rapid cortical reorganization occur after cortical injuries (Nishibe et al. 2015; Nishibe et al. 2010). Our recent finding that the cortical remodeling occurred in a lesser extent in HRM than in WT through reach-to-grasp training (Nishibe submitted) indicates Reelin haploinsufficiny causes consequences to neuroanatomical substrates, affecting the synaptic remodeling within the cortical network. The results, hence, indicate that Reelin haploinsuffi-ciency may result in a lesser extent of cortical remodeling, possibly following a brain injury as well (i.e., during re-shaping and recovery of behavioral function).

Various reports suggest the reduced expression level of Reelin affects the number of dendrites in the prefrontal cortex, hippocampus, and cerebellum (Yabut et al. 2007; Nichols and Olson 2010; O'Dell et al. 2015). In brain slice culture experiments, addition of Reelin interfering antibodies, receptor antagonists, and Dab1 phosphorylation inhibitors all prevented dendrite outgrowth in the hippocampus. Whereas, addition of recombinant Reelin protein rescued the dendritic deficit in *reeler* brain cultures (Niu et al. 2004). Extensive remodeling of dendrites within the motor cortex has been observed using two-photon microscopy during reach-to-grasp learning in mice (Xu et al. 2009; Roth et al. 2020). Thus, it is likely that Reelin plays a role in repair of neuronal connections after brain injury by accelerating dendrogenesis of the cortical neurons.

CONCLUSION

Reelin is required for corticogenesis during brain development in mammals. *RELN* has also been suggested to be one of the candidate genes responsible for some of both psychiatric and motor symptoms observed in neurological and psychiatric diseases. Recent studies focusing on the cerebral cortex using the *reeler* model showed that Reelin is involved in the regulation of synaptic and dendritic functions as well as their plasticity in adulthood. We propose that the differential regulatory function of Reelin is observed under the influence of developmental abnormalities and in typical brain development. Further investigations on the function of Reelin in the cerebral cortex will provide us with information for finding a clinical treatment target for patients with developmental disorders or with brain injury who experience motor-related disability.

REFERENCES

Ammassari-Teule, M., C. Sgobio, F. Biamonte, C. Marrone, N. B. Mercuri, and F. Keller. 2009. "Reelin haploinsufficiency reduces the density of PV+ neurons in circumscribed regions of the striatum and selectively alters striatal-based behaviors." *Psychopharmacology (Berl)* 204 (3): 511-21. https://doi.org/10.1007/s00213-009-1483-x.

Barr, A. M., K. N. Fish, A. Markou, and W. G. Honer. 2008. "Heterozygous reeler mice exhibit alterations in sensorimotor gating but not presynaptic proteins." *Eur J Neurosci* 27 (10): 2568-74. https://doi.org/10.1111/j.1460-9568.2008.06233.x.

Beffert, U., E. J. Weeber, A. Durudas, S. Qiu, I. Masiulis, J. D. Sweatt, W. P. Li, G. Adelmann, M. Frotscher, R. E. Hammer, and J. Herz. 2005. "Modulation of synaptic plasticity and memory by Reelin involves differential splicing of the lipoprotein receptor Apoer2." *Neuron* 47 (4): 567-79. https://doi.org/10.1016/j.neuron.2005.07.007.

Boycott, K. M., S. Flavelle, A. Bureau, H. C. Glass, T. M. Fujiwara, E. Wirrell, K. Davey, A. E. Chudley, J. N. Scott, D. R. McLeod, and J. S. Parboosingh. 2005. "Homozygous deletion of the very low density lipoprotein receptor gene causes autosomal recessive cerebellar hypoplasia with cerebral gyral simplification." *Am J Hum Genet* 77 (3): 477-83. https://doi.org/10.1086/444400.

Brigman, J. L., K. E. Padukiewicz, M. L. Sutherland, and L. A. Rothblat. 2006. "Executive functions in the heterozygous reeler mouse model of schizophrenia." *Behav Neurosci* 120 (4): 984-8. https://doi.org/10.1037/0735-7044.120.4.984.

Carboni, G., P. Tueting, L. Tremolizzo, I. Sugaya, J. Davis, E. Costa, and A. Guidotti. 2004. "Enhanced dizocilpine efficacy in heterozygous reeler mice relates to GABA turnover downregulation." *Neuropharmacology* 46 (8): 1070-81. https://doi.org/10.1016/j.neuropharm.2004.02.001.

Caviness, V. S., Jr., and P. Rakic. 1978. "Mechanisms of cortical development: a view from mutations in mice." *Annu Rev Neurosci* 1: 297-326. https://doi.org/10.1146/annurev.ne.01.030178.001501.

Caviness, V. S., Jr., D. K. So, and R. L. Sidman. 1972. "The hybrid reeler mouse." *J Hered* 63 (5): 241-6. https://doi.org/10.1093/oxfordjournals.jhered.a108286.

Chen, Y., U. Beffert, M. Ertunc, T. S. Tang, E. T. Kavalali, I. Bezprozvanny, and J. Herz. 2005. "Reelin modulates NMDA receptor activity in cortical neurons." *J Neurosci* 25 (36): 8209-16. https://doi.org/10.1523/jneurosci.1951-05.2005.

Costa, E., Y. Chen, J. Davis, E. Dong, J. S. Noh, L. Tremolizzo, M. Veldic, D. R. Grayson, and A. Guidotti. 2002. "REELIN and schizophrenia: a disease at the interface of the genome and the epigenome." *Mol Interv* 2 (1): 47-57. https://doi.org/10.1124/mi.2.1.47.

Courtes, S., J. Vernerey, L. Pujadas, K. Magalon, H. Cremer, E. Soriano, P. Durbec, and M. Cayre. 2011. "Reelin controls progenitor cell migration in the healthy and pathological adult mouse brain." *PLoS One* 6 (5): e20430. https://doi.org/10.1371/journal.pone.0020430.

D'Arcangelo, G., G. G. Miao, S. C. Chen, H. D. Soares, J. I. Morgan, and T. Curran. 1995. "A protein related to extracellular matrix proteins deleted in the mouse mutant reeler." *Nature* 374 (6524): 719-23. https://doi.org/10.1038/374719a0.

Devisme, L., C. Bouchet, M. Gonzales, E. Alanio, A. Bazin, B. Bessieres, N. Bigi, P. Blanchet, D. Bonneau, M. Bonnieres, M. Bucourt, D. Carles, B. Clarisse, S. Delahaye, C. Fallet-Bianco, D. Figarella-Branger, D. Gaillard, B. Gasser, A. L. Delezoide, F. Guimiot, M. Joubert, N. Laurent, A. Laquerriere, A. Liprandi, P. Loget, P. Marcorelles, J. Martinovic, F. Menez, S. Patrier, F. Pelluard, M. J. Perez, C. Rouleau, S. Triau, T. Attie-Bitach, S. Vuillaumier-Barrot, N. Seta, and F. Encha-Razavi. 2012. "Cobblestone lissencephaly: neuropathological subtypes and correlations with genes of dystroglycanopathies." *Brain* 135 (Pt 2): 469-82. https://doi.org/10.1093/brain/awr357.

Diener, H. C., J. Dichgans, B. Guschlbauer, M. Bacher, H. Rapp, and T. Klockgether. 1992. "The coordination of posture and voluntary movement in patients with cerebellar dysfunction." *Mov Disord* 7 (1): 14-22. https://doi.org/10.1002/mds.870070104.

Exner, C., G. Weniger, C. Schmidt-Samoa, and E. Irle. 2006. "Reduced size of the pre-supplementary motor cortex and impaired motor sequence learning in first-episode schizophrenia." *Schizophr Res* 84 (2-3): 386-96. https://doi.org/10.1016/j.schres.2006.03.013.

Falconer, D. S. 1951. "Two new mutants, 'trembler' and 'reeler', with neurological actions in the house mouse (Mus musculus L.)." *J Genet* 50 (2): 192-201.

Fatemi, S. H., and T. D. Folsom. 2011. "The role of fragile X mental retardation protein in major mental disorders." *Neuropharmacology* 60 (7-8): 1221-6. https://doi.org/10.1016/j.neuropharm.2010.11.011.

Fatemi, S. H., J. M. Stary, J. A. Earle, M. Araghi-Niknam, and E. Eagan. 2005. "GABAergic dysfunction in schizophrenia and mood disorders as reflected by decreased levels of glutamic acid decarboxylase 65 and 67 kDa and Reelin proteins in cerebellum." *Schizophr Res* 72 (2-3): 109-22. https://doi.org/10.1016/j.schres.2004.02.017.

Feng, L., N. S. Allen, S. Simo, and J. A. Cooper. 2007. "Cullin 5 regulates Dab1 protein levels and neuron positioning during cortical development." *Genes Dev* 21 (21): 2717-30. https://doi.org/10.1101/gad.1604207.

Gowert, N. S., I. Kruger, M. Klier, L. Donner, F. Kipkeew, M. Gliem, N. J. Bradshaw, D. Lutz, S. Kober, H. Langer, S. Jander, K. Jurk, M. Frotscher, C. Korth, H. H. Bock, and M. Elvers. 2017. "Loss of Reelin protects mice against arterial thrombosis by impairing integrin activation and thrombus formation under high shear conditions." *Cell Signal* 40: 210-221. https://doi.org/10.1016/j.cellsig.2017.09.016.

Gschwandtner, U., M. Pfluger, J. Aston, S. Borgwardt, M. Drewe, R. D. Stieglitz, and A. Riecher-Rossler. 2006. "Fine motor function and neuropsychological deficits in individuals at risk for schizophrenia." *Eur Arch Psychiatry Clin Neurosci* 256 (4): 201-6. https://doi.org/10.1007/s00406-005-0626-2.

Guerrini, R., and W. B. Dobyns. 2014. "Malformations of cortical development: clinical features and genetic causes." *Lancet Neurol* 13 (7): 710-26. https://doi.org/10.1016/s1474-4422(14)70040-7.

Guidotti, A., J. Auta, J. M. Davis, V. Di-Giorgi-Gerevini, Y. Dwivedi, D. R. Grayson, F. Impagnatiello, G. Pandey, C. Pesold, R. Sharma, D. Uzunov, and E. Costa. 2000. "Decrease in reelin and glutamic acid decarboxylase67 (GAD67) expression in schizophrenia and bipolar disorder: a postmortem brain study." *Arch Gen Psychiatry* 57 (11): 1061-9.

Hallett, M. 2001. "Plasticity of the human motor cortex and recovery from stroke." *Brain Res Brain Res Rev* 36 (2-3): 169-74.

Hashimoto, T., D. Arion, T. Unger, J. G. Maldonado-Aviles, H. M. Morris, D. W. Volk, K. Mirnics, and D. A. Lewis. 2008. "Alterations in GABA-related transcriptome in the dorsolateral prefrontal cortex of subjects with schizophrenia." *Mol Psychiatry* 13 (2): 147-61. https://doi.org/10.1038/sj.mp.4002011.

Hong, S. E., Y. Y. Shugart, D. T. Huang, S. A. Shahwan, P. E. Grant, J. O. Hourihane, N. D. Martin, and C. A. Walsh. 2000. "Autosomal recessive lissencephaly with cerebellar hypoplasia is associated with human RELN mutations." *Nat Genet* 26 (1): 93-6. https://doi.org/10.1038/79246.

Hourihane, J. O., C. P. Bennett, R. Chaudhuri, S. A. Robb, and N. D. Martin. 1993. "A sibship with a neuronal migration defect, cerebellar hypoplasia and congenital lymphedema." *Neuropediatrics* 24 (1): 43-6. https://doi.org/10.1055/s-2008-1071511.

Howell, B. W., T. M. Herrick, J. D. Hildebrand, Y. Zhang, and J. A. Cooper. 2000. "Dab1 tyrosine phosphorylation sites relay positional signals during mouse brain development." *Curr Biol* 10 (15): 877-85. https://doi.org/10.1016/s0960-9822(00)00608-4.

Imai, H., H. Shoji, M. Ogata, Y. Kagawa, Y. Owada, T. Miyakawa, K. Sakimura, T. Terashima, and Y. Katsuyama. 2017. "Dorsal Forebrain-Specific Deficiency of Reelin-Dab1 Signal Causes Behavioral Abnormalities Related to Psychiatric Disorders." *Cereb Cortex* 27 (7): 3485-3501. https://doi.org/10.1093/cercor/bhv334.

Impagnatiello, F., A. R. Guidotti, C. Pesold, Y. Dwivedi, H. Caruncho, M. G. Pisu, D. P. Uzunov, N. R. Smalheiser, J. M. Davis, G. N. Pandey, G. D. Pappas, P. Tueting, R. P. Sharma, and E. Costa. 1998. "A decrease of reelin expression as a putative vulnerability factor in schizophrenia." *Proc Natl Acad Sci U S A* 95 (26): 15718-23.

Jogems-Kosterman, B. J., W. Hulstijn, E. Wezenberg, and J. J. van Hoof. 2006. "Movement planning deficits in schizophrenia: failure to inhibit automatic response tendencies." *Cogn Neuropsychiatry* 11 (1): 47-64. https://doi.org/10.1080/13546800444000173.

Jones, T. A., R. P. Allred, S. C. Jefferson, A. L. Kerr, D. A. Woodie, S. Y. Cheng, and D. L. Adkins. 2013. "Motor system plasticity in stroke models: intrinsically use-dependent, unreliably useful." *Stroke* 44 (6 Suppl 1): S104-6. https://doi.org/10.1161/strokeaha.111.000037.

Kern, R. S., M. F. Green, B. D. Marshall, Jr., W. C. Wirshing, D. Wirshing, S. McGurk, S. R. Marder, and J. Mintz. 1998. "Risperidone vs. haloperidol on reaction time, manual dexterity, and motor learning in treatment-resistant schizophrenia patients." *Biol Psychiatry* 44 (8): 726-32.

Kolb, L. E., Z. Arlier, C. Yalcinkaya, A. K. Ozturk, J. A. Moliterno, O. Erturk, F. Bayrakli, B. Korkmaz, M. L. DiLuna, K. Yasuno, K. Bilguvar, T. Ozcelik, B. Tuysuz, M. W. State, and M. Gunel. 2010. "Novel VLDLR microdeletion identified in two Turkish siblings with pachygyria and pontocerebellar atrophy." *Neurogenetics* 11 (3): 319-25. https://doi.org/10.1007/s10048-009-0232-y.

Krueger, D. D., J. L. Howell, B. F. Hebert, P. Olausson, J. R. Taylor, and A. C. Nairn. 2006. "Assessment of cognitive function in the heterozygous reeler mouse." *Psychopharmacology (Berl)* 189 (1): 95-104. https://doi.org/10.1007/s00213-006-0530-0.

Kutiyanawalla, A., W. Promsote, A. Terry, and A. Pillai. 2012. "Cysteamine treatment ameliorates alterations in GAD67 expression and spatial memory in heterozygous reeler mice." *Int J Neuropsychopharmacol* 15 (8): 1073-86. https://doi.org/10.1017/s1461145711001180.

Lambert de Rouvroit, C., V. de Bergeyck, C. Cortvrindt, I. Bar, Y. Eeckhout, and A. M. Goffinet. 1999. "Reelin, the extracellular matrix protein deficient in reeler mutant mice, is processed by a metalloproteinase." *Exp Neurol* 156 (1): 214-7. https://doi.org/10.1006/exnr.1998.7007.

Liu, W. S., C. Pesold, M. A. Rodriguez, G. Carboni, J. Auta, P. Lacor, J. Larson, B. G. Condie, A. Guidotti, and E. Costa. 2001. "Downregulation of dendritic spine and glutamic acid decarboxylase 67 expressions in the reelin haploinsufficient heterozygous reeler mouse." *Proc Natl Acad Sci U S A* 98 (6): 3477-82. https://doi.org/10.1073/pnas.051614698.

Massalini, S., S. Pellegatta, F. Pisati, G. Finocchiaro, M. G. Farace, and S. A. Ciafre. 2009. "Reelin affects chain-migration and differentiation of neural precursor cells." *Mol Cell Neurosci* 42 (4): 341-9. https://doi.org/10.1016/j.mcn.2009.08.006.

McAlonan, G. M., E. Daly, V. Kumari, H. D. Critchley, T. van Amelsvoort, J. Suckling, A. Simmons, T. Sigmundsson, K. Greenwood, A. Russell, N. Schmitz, F. Happe, P. Howlin, and D. G. Murphy. 2002. "Brain anatomy and sensorimotor gating in Asperger's syndrome." *Brain* 125 (Pt 7): 1594-606. https://doi.org/10.1093/brain/awf150.

Midorikawa, A., R. Hashimoto, H. Noguchi, O. Saitoh, H. Kunugi, and K. Nakamura. 2008. "Impairment of motor dexterity in schizophrenia assessed by a novel finger movement test." *Psychiatry Res* 159 (3): 281-9. https://doi.org/10.1016/j.psychres.2007.04.004.

Mill, J., T. Tang, Z. Kaminsky, T. Khare, S. Yazdanpanah, L. Bouchard, P. Jia, A. Assadzadeh, J. Flanagan, A. Schumacher, S. C. Wang, and A. Petronis. 2008. "Epigenomic profiling reveals DNA-methylation changes associated with major psychosis." *Am J Hum Genet* 82 (3): 696-711. https://doi.org/10.1016/j.ajhg.2008.01.008.

Mirkowski, M., A. McIntyre, P. Faltynek, N. Sequeira, C. Cassidy, and R. Teasell. 2019. "Nonpharmacological rehabilitation interventions for motor and cognitive outcomes following pediatric stroke: a systematic review." *Eur J Pediatr* 178 (4): 433-454. https://doi.org/10.1007/s00431-019-03350-7.

Morrens, M., L. Docx, and S. Walther. 2014. "Beyond boundaries: in search of an integrative view on motor symptoms in schizophrenia." *Front Psychiatry* 5: 145. https://doi.org/10.3389/fpsyt.2014.00145.

Myers, W. A. 1970. "Some observations on "reeler", a neuromuscular mutation in mice." *Behav Genet* 1 (3): 225-34. https://doi.org/10.1007/bf01074654.

Nakano, Y., T. Kohno, T. Hibi, S. Kohno, A. Baba, K. Mikoshiba, K. Nakajima, and M. Hattori. 2007. "The extremely conserved C-terminal region of Reelin is not necessary for secretion but is required for efficient activation of downstream signaling." *J Biol Chem* 282 (28): 20544-52. https://doi.org/10.1074/jbc.M702300200.

Nichols, A. J., and E. C. Olson. 2010. "Reelin promotes neuronal orientation and dendritogenesis during preplate splitting." *Cereb Cortex* 20 (9): 2213-23. https://doi.org/10.1093/cercor/bhp303.

Nishibe, M., S. Barbay, D. Guggenmos, and R. J. Nudo. 2010. "Reorganization of motor cortex after controlled cortical impact in rats and implications for functional recovery." *Journal of Neurotrauma* 27 (12): 2221-32. https://doi.org/10.1089/neu.2010.1456.

Nishibe, M., E. T. Urban, 3rd, S. Barbay, and R. J. Nudo. 2015. "Rehabilitative training promotes rapid motor recovery but delayed motor map reorganization in a rat cortical ischemic infarct model." *Neurorehabil Neural Repair* 29 (5): 472-82. https://doi.org/10.1177/1545968314543499.

Niu, S., A. Renfro, C. C. Quattrocchi, M. Sheldon, and G. D'Arcangelo. 2004. "Reelin promotes hippocampal dendrite development through the VLDLR/ApoER2-Dab1 pathway." *Neuron* 41 (1): 71-84.

Northoff, G., D. F. Braus, A. Sartorius, D. Khoram-Sefat, M. Russ, J. Eckert, M. Herrig, A. Leschinger, B. Bogerts, and F. A. Henn. 1999. "Reduced activation and altered laterality in two neuroleptic-naive catatonic patients during a motor task in functional MRI." *Psychol Med* 29 (4): 997-1002.

Northoff, G., R. Steinke, C. Czcervenka, R. Krause, S. Ulrich, P. Danos, D. Kropf, H. Otto, and B. Bogerts. 1999. "Decreased density of GABA-A receptors in the left sensorimotor cortex in akinetic catatonia: investigation of in vivo benzodiazepine receptor binding." *J Neurol Neurosurg Psychiatry* 67 (4): 445-50.

Nudo, R. J., E. J. Plautz, and S. B. Frost. 2001. "Role of adaptive plasticity in recovery of function after damage to motor cortex." *Muscle Nerve* 24 (8): 1000-19. https://doi.org/10.1002/mus.1104 [pii].

Nullmeier, S., P. Panther, H. Dobrowolny, M. Frotscher, S. Zhao, H. Schwegler, and R. Wolf. 2011. "Region-specific alteration of GABAergic markers in the brain of heterozygous reeler mice." *Eur J Neurosci* 33 (4): 689-98. https://doi.org/10.1111/j.1460-9568.2010.07563.x.

O'Dell, R. S., D. A. Cameron, W. R. Zipfel, and E. C. Olson. 2015. "Reelin Prevents Apical Neurite Retraction during Terminal Translocation and Dendrite Initiation." *J Neurosci* 35 (30): 10659-74. https://doi.org/10.1523/jneurosci.1629-15.2015.

Ognibene, E., W. Adriani, O. Granstrem, S. Pieretti, and G. Laviola. 2007. "Impulsivity-anxiety-related behavior and profiles of morphine-induced analgesia in heterozygous reeler mice." *Brain Res* 1131 (1): 173-80. https://doi.org/10.1016/j.brainres.2006.11.007.

Patel, S., and A. J. Barkovich. 2002. "Analysis and classification of cerebellar malformations." *AJNR Am J Neuroradiol* 23 (7): 1074-87.

Payoux, P., K. Boulanouar, C. Sarramon, N. Fabre, S. Descombes, M. Galitsky, C. Thalamas, C. Brefel-Courbon, U. Sabatini, C. Manelfe, F. Chollet, L. Schmitt, and O. Rascol. 2004. "Cortical motor activation in akinetic schizophrenic patients: a pilot functional MRI study." *Mov Disord* 19 (1): 83-90. https://doi.org/10.1002/mds.10598.

Perry, W., M. A. Geyer, and D. L. Braff. 1999. "Sensorimotor gating and thought disturbance measured in close temporal proximity in schizophrenic patients." *Arch Gen Psychiatry* 56 (3): 277-81. https://doi.org/10.1001/archpsyc.56.3.277.

Qiu, S., K. M. Korwek, A. R. Pratt-Davis, M. Peters, M. Y. Bergman, and E. J. Weeber. 2006. "Cognitive disruption and altered hippocampus synaptic function in Reelin haploinsufficient mice." *Neurobiol Learn Mem* 85 (3): 228-42. https://doi.org/10.1016/j.nlm.2005.11.001.

Quattrocchi, C. C., F. Wannenes, A. M. Persico, S. A. Ciafre, G. D'Arcangelo, M. G. Farace, and F. Keller. 2002. "Reelin is a serine protease of the extracellular matrix." *J Biol Chem* 277 (1): 303-9. https://doi.org/10.1074/jbc.M106996200.

Rogers, J. T., I. Rusiana, J. Trotter, L. Zhao, E. Donaldson, D. T. Pak, L. W. Babus, M. Peters, J. L. Banko, P. Chavis, G. W. Rebeck, H. S. Hoe, and E. J. Weeber. 2011. "Reelin supplementation enhances cognitive ability, synaptic plasticity, and dendritic spine density." *Learn Mem* 18 (9): 558-64. https://doi.org/10.1101/lm.2153511.

Ross, M. E., K. Swanson, and W. B. Dobyns. 2001. "Lissencephaly with cerebellar hypoplasia (LCH): a heterogeneous group of cortical malformations." *Neuropediatrics* 32 (5): 256-63. https://doi.org/10.1055/s-2001-19120.

Roth, R. H., R. H. Cudmore, H. L. Tan, I. Hong, Y. Zhang, and R. L. Huganir. 2020. "Cortical Synaptic AMPA Receptor Plasticity during Motor Learning." *Neuron* 105 (5): 895-908 e5. https://doi.org/10.1016/j.neuron.2019.12.005.

Royaux, I., C. Lambert de Rouvroit, G. D'Arcangelo, D. Demirov, and A. M. Goffinet. 1997. "Genomic organization of the mouse reelin gene." *Genomics* 46 (2): 240-50. https://doi.org/10.1006/geno.1997.4983.

Salinger, W. L., P. Ladrow, and C. Wheeler. 2003. "Behavioral phenotype of the reeler mutant mouse: effects of RELN gene dosage and social isolation." *Behav Neurosci* 117 (6): 1257-75. https://doi.org/10.1037/0735-7044.117.6.1257.

Scheuerecker, J., S. Ufer, M. Kapernick, M. Wiesmann, H. Bruckmann, E. Kraft, D. Seifert, N. Koutsouleris, H. J. Moller, and E. M. Meisenzahl. 2009. "Cerebral network deficits in post-acute catatonic schizophrenic patients measured by fMRI." *J Psychiatr Res* 43 (6): 607-14. https://doi.org/10.1016/j.jpsychires.2008.08.005.

Schroder, J., M. Essig, K. Baudendistel, T. Jahn, I. Gerdsen, A. Stockert, L. R. Schad, and M. V. Knopp. 1999. "Motor dysfunction and sensorimotor cortex activation changes in schizophrenia: A study with functional magnetic resonance imaging." *Neuroimage* 9 (1): 81-7. https://doi.org/10.1006/nimg.1998.0387.

Schroder, J., F. Wenz, L. R. Schad, K. Baudendistel, and M. V. Knopp. 1995. "Sensorimotor cortex and supplementary motor area changes in schizophrenia. A study with functional magnetic resonance imaging." *Br J Psychiatry* 167 (2): 197-201.

Schwartz, F., A. Carr, R. Munich, E. Bartuch, B. Lesser, D. Rescigno, and B. Viegener. 1990. "Voluntary motor performance in psychotic disorders: a replication study." *Psychol Rep* 66 (3 Pt 2): 1223-34. https://doi.org/10.2466/pr0.1990.66.3c.1223.

Sheldon, M., D. S. Rice, G. D'Arcangelo, H. Yoneshima, K. Nakajima, K. Mikoshiba, B. W. Howell, J. A. Cooper, D. Goldowitz, and T. Curran. 1997. "Scrambler and yotari disrupt the disabled gene and produce a reeler-like phenotype in mice." *Nature* 389 (6652): 730-3. https://doi.org/10.1038/39601.

Silva, L. R., M. J. Gutnick, and B. W. Connors. 1991. "Laminar distribution of neuronal membrane properties in neocortex of normal and reeler mouse." *J Neurophysiol* 66 (6): 2034-40.

Stary, C. M., L. Xu, X. Sun, Y. B. Ouyang, R. E. White, J. Leong, J. Li, X. Xiong, and R. G. Giffard. 2015. "MicroRNA-200c contributes to injury from transient focal cerebral ischemia by targeting Reelin." *Stroke* 46 (2): 551-6. https://doi.org/10.1161/strokeaha.114.007041.

Tan, A. P., W. K. Chong, and K. Mankad. 2018. "Comprehensive genotype-phenotype correlation in lissencephaly." *Quant Imaging Med Surg* 8 (7): 673-693. https://doi.org/10.21037/qims.2018.08.08.

Teixeira, C. M., E. D. Martin, I. Sahun, N. Masachs, L. Pujadas, A. Corvelo, C. Bosch, D. Rossi, A. Martinez, R. Maldonado, M. Dierssen, and E. Soriano. 2011. "Overexpression of Reelin prevents the manifestation of behavioral phenotypes related to schizophrenia and bipolar disorder." *Neuropsychopharmacology* 36 (12): 2395-405. https://doi.org/10.1038/npp.2011.153.

Terashima, T. 1995a. "Anatomy, development and lesion-induced plasticity of rodent corticospinal tract." *Neurosci Res* 22 (2): 139-61.

---. 1995b. "Course and collaterals of corticospinal fibers arising from the sensorimotor cortex of the reeler mouse." *Dev Neurosci* 17 (1): 8-19.

Terashima, T., C. Takayama, R. Ichikawa, and Y. Inoue. 1992. "Dendritic arborization of large pyramidal neurons in the motor cortex of normal and reeler mutant mouse." *Okajimas Folia Anat Jpn* 68 (6): 351-63.

Teremetz, M., I. Amado, N. Bendjemaa, M. O. Krebs, P. G. Lindberg, and M. A. Maier. 2014. "Deficient grip force control in schizophrenia: behavioral and modeling evidence for altered motor inhibition and motor noise." *PLoS One* 9 (11): e111853. https://doi.org/10.1371/journal.pone.0111853.

Teremetz, M., L. Carment, L. Brenugat-Herne, M. Croca, J. P. Bleton, M. O. Krebs, M. A. Maier, I. Amado, and P. G. Lindberg. 2017. "Manual Dexterity in Schizophrenia-A Neglected Clinical Marker?" *Front Psychiatry* 8: 120. https://doi.org/10.3389/fpsyt.2017.00120.

Tochigi, M., K. Iwamoto, M. Bundo, A. Komori, T. Sasaki, N. Kato, and T. Kato. 2008. "Methylation status of the reelin promoter region in the brain of schizophrenic patients." *Biol Psychiatry* 63 (5): 530-3. https://doi.org/10.1016/j.biopsych.2007.07.003.

Trommsdorff, M., M. Gotthardt, T. Hiesberger, J. Shelton, W. Stockinger, J. Nimpf, R. E. Hammer, J. A. Richardson, and J. Herz. 1999. "Reeler/Disabled-like disruption of neuronal migration in knockout mice lacking the VLDL receptor and ApoE receptor 2." *Cell* 97 (6): 689-701. https://doi.org/10.1016/s0092-8674(00)80782-5.

Trotter, J., G. H. Lee, T. M. Kazdoba, B. Crowell, J. Domogauer, H. M. Mahoney, S. J. Franco, U. Muller, E. J. Weeber, and G. D'Arcangelo. 2013. "Dab1 is required for synaptic plasticity and associative learning." *J Neurosci* 33 (39): 15652-68. https://doi.org/10.1523/jneurosci.2010-13.2013.

Tueting, P., E. Costa, Y. Dwivedi, A. Guidotti, F. Impagnatiello, R. Manev, and C. Pesold. 1999. "The phenotypic characteristics of heterozygous reeler mouse." *Neuroreport* 10 (6): 1329-34.

Turetsky, B. I., M. E. Calkins, G. A. Light, A. Olincy, A. D. Radant, and N. R. Swerdlow. 2007. "Neurophysiological endophenotypes of schizophrenia: the viability of selected candidate measures." *Schizophr Bull* 33 (1): 69-94. https://doi.org/10.1093/schbul/sbl060.

Ware, M. L., J. W. Fox, J. L. Gonzalez, N. M. Davis, C. Lambert de Rouvroit, C. J. Russo, S. C. Chua, Jr., A. M. Goffinet, and C. A. Walsh. 1997. "Aberrant splicing of a mouse disabled homolog, mdab1, in the scrambler mouse." *Neuron* 19 (2): 239-49. https://doi.org/10.1016/s0896-6273(00)80936-8.

Weeber, E. J., U. Beffert, C. Jones, J. M. Christian, E. Forster, J. D. Sweatt, and J. Herz. 2002. "Reelin and ApoE receptors cooperate to enhance hippocampal synaptic plasticity and learning." *J Biol Chem* 277 (42): 39944-52. https://doi.org/10.1074/jbc.M205147200.

Won, S. J., S. H. Kim, L. Xie, Y. Wang, X. O. Mao, K. Jin, and D. A. Greenberg. 2006. "Reelin-deficient mice show impaired neurogenesis and increased stroke size." *Exp Neurol* 198 (1): 250-9. https://doi.org/10.1016/j.expneurol.2005.12.008.

Xu, T., X. Yu, A. J. Perlik, W. F. Tobin, J. A. Zweig, K. Tennant, T. Jones, and Y. Zuo. 2009. "Rapid formation and selective stabilization of synapses for enduring motor memories." *Nature* 462 (7275): 915-9. https://doi.org/10.1038/nature08389.

Yabut, O., A. Renfro, S. Niu, J. W. Swann, O. Marin, and G. D'Arcangelo. 2007. "Abnormal laminar position and dendrite development of interneurons in the reeler forebrain." *Brain Res* 1140: 75-83. https://doi.org/10.1016/j.brainres.2005.09.070.

Yamamoto, T., T. Setsu, A. Okuyama-Yamamoto, and T. Terashima. 2009. "Histological study in the brain of the reelin/Dab1-compound mutant mouse." *Anat Sci Int* 84 (3): 200-9. https://doi.org/10.1007/s12565-008-0009-7.

Zaki, M., M. Shehab, A. A. El-Aleem, G. Abdel-Salam, H. B. Koeller, Y. Ilkin, M. E. Ross, W. B. Dobyns, and J. G. Gleeson. 2007. "Identification of a novel recessive RELN mutation using a homozygous balanced reciprocal translocation." *Am J Med Genet A* 143a (9): 939-44. https://doi.org/10.1002/ajmg.a.31667.

INDEX

A

acetylcholine, 62, 69, 79, 80, 81, 82, 83, 85, 86, 87, 88, 89, 90, 91, 92, 93
acetylcholinesterase, 72, 86, 91
ACh, 62, 65, 67, 68, 69, 70, 71, 72, 73, 74, 75, 76, 78
acid, 21, 52, 83, 92
adaptation, 73, 87, 96, 109
adhesion, 11, 15, 26, 28
adulthood, 22, 110, 111
allele, 36, 104, 107
Alzheimer, 62, 77, 80, 82, 85, 86, 87, 89, 91, 92, 95, 96, 97
amphibia, 5, 6, 12, 17, 25, 29, 32
anatomy, 63, 86, 117
apoptosis, 5, 13, 48
arousal, viii, 62, 65, 68, 69, 71, 75, 78
ataxia, viii, 99, 104, 107
atrophy, 80, 90, 116
avian, 5, 27, 49, 53
axons, 69, 73, 96

B

basal forebrain, vii, viii, 61, 62, 70, 79, 80, 81, 82, 83, 84, 86, 87, 88, 89, 90, 92, 93, 96
basal nucleus of Meynert, 63, 64
behaviors, 27, 29, 101, 112
bipolar disorder, 107, 115, 121
brain, vii, viii, 1, 2, 3, 4, 6, 9, 11, 20, 23, 31, 34, 35, 36, 37, 38, 44, 46, 47, 48, 49, 52, 53, 54, 56, 58, 63, 64, 70, 71, 72, 73, 81, 85, 86, 93, 95, 99, 100, 102, 110, 111, 113, 115, 117, 119, 122, 123
brain injury, 100, 110, 111

C

cell fate, 5, 40, 44, 55
cell line, 4, 11, 55
cell surface, 22, 26, 27
central nervous system, 5, 39, 40, 53, 55, 81
cerebellum, 104, 106, 107, 111, 114

cerebral cortex, viii, 2, 62, 69, 77, 78, 80, 82, 85, 86, 87, 99, 100, 102, 103, 104, 105, 106, 107, 111
chicken, 9, 10, 12, 13, 23, 30, 33, 34, 37
cholesterol, 21, 22, 40, 56
choline, 73, 77, 85, 87, 89, 94, 107
cilia, viii, 2, 3, 22, 23, 36, 40, 43, 45, 49, 51, 57
cilium, 10, 15, 21, 22, 23, 46
classification, 63, 64, 119
cleavage, 21, 23, 53
CNS, 6, 50, 82
cognition, 79, 85, 109
cognitive function, viii, 61, 62, 81, 105, 116
cognitive functions, viii, 61, 62, 81, 90
cognitive impairment, 77, 80, 86, 105
compartment, vii, 1, 2, 3, 7, 10, 11, 15, 24, 25, 26, 27, 28, 29, 41, 58
connectivity, 55, 64, 75, 93, 106
correlation(s), 83, 88, 104, 113, 121
cortex, viii, 2, 61, 62, 64, 65, 66, 67, 68, 69, 70, 72, 73, 74, 75, 76, 77, 78, 80, 82, 83, 85, 86, 87, 88, 89, 90, 91, 92, 93, 94, 95, 99, 100, 102, 103, 104, 105, 106, 107, 108, 111, 114, 115, 118, 119, 120, 121
cortical malformation, 104, 106, 120
cortical neurons, 65, 68, 73, 74, 76, 77, 79, 96, 100, 111, 113
cues, vii, viii, 1, 3, 4, 12, 17, 21, 30, 34, 37, 84, 85

D

defects, viii, 2, 3, 18, 31, 32, 34, 35, 36, 104
deficit, 34, 110, 111
degradation, 12, 15, 102
detection, 73, 74, 76, 84, 89, 90, 92, 96
developmental biology, 2
developmental disorder, 38, 104, 111
diagonal band of Broca, 63, 66, 70
diseases, 80, 100, 111

disorder, 34, 107, 109
distribution, 65, 68, 81, 87, 121
Drosophila, 21, 22, 24, 28, 41, 57, 102

E

EEG, 62, 69, 70, 71, 82, 83, 88, 94
embryogenesis, 2, 23
encoding, 17, 39, 72
etiology, viii, 2, 3, 35, 37
evidence, viii, 1, 22, 76, 78, 83, 85, 122
evolution, 5, 48, 89
excitability, 65, 72, 73
extracellular matrix, 26, 101, 113, 116, 120

F

fiber(s), 64, 74, 85, 121
forebrain, v, vii, viii, 1, 2, 3, 4, 6, 7, 8, 11, 12, 17, 18, 19, 21, 23, 25, 26, 27, 28, 30, 31, 34, 35, 36, 37, 39, 40, 42, 45, 47, 48, 49, 50, 51, 52, 53, 54, 57, 58, 60, 61, 62, 63, 65, 70, 77, 79, 80, 81, 82, 83, 84, 85, 86, 87, 88, 89, 90, 91, 92, 93, 94, 96, 105, 115, 123
formation, viii, ix, 1, 3, 11, 12, 13, 15, 16, 18, 20, 23, 24, 25, 26, 31, 33, 34, 35, 37, 39, 41, 43, 45, 49, 55, 56, 57, 58, 60, 64, 88, 100, 103, 106, 114, 123
France, 1, 38, 58, 59
frontal cortex, 64, 89, 91
functional MRI, 109, 118, 119

G

GABA, 70, 91, 92, 108, 112, 115, 118
GABAergic, 54, 62, 63, 68, 69, 70, 76, 84, 85, 87, 108, 114, 119
gamma oscillations, 71, 76, 80, 90, 91
gastrulation, 6, 47, 55

Index

gene expression, 4, 7, 9, 12, 15, 29, 50, 53, 56
genes, 4, 5, 15, 17, 23, 24, 28, 30, 31, 34, 35, 36, 38, 41, 42, 45, 46, 52, 54, 55, 102, 104, 107, 111, 113
glutamic acid, 108, 114, 115, 117
growth, vii, 1, 3, 5, 27, 34, 37, 48, 50, 101, 104

H

hippocampus, viii, 62, 63, 64, 68, 70, 71, 77, 79, 80, 90, 92, 95, 100, 102, 104, 107, 108, 111, 119
holoprosencephaly, 2, 34, 44, 45, 46, 50, 51, 52
human, viii, 19, 35, 36, 38, 42, 52, 63, 67, 68, 88, 89, 92, 99, 104, 106, 115
human brain, viii, 63, 89, 99
human genome, viii, 19, 99
hypoplasia, 104, 107, 112, 115, 120
hypothalamus, 11, 31, 45
hypothesis, 20, 33, 35, 77, 82, 90, 91
hypotonia, 104, 107

I

identity, 5, 24, 25, 29, 33, 37, 44, 47, 48, 52
impairments, vii, viii, 99, 100, 105, 106, 109, 110
in vivo, 10, 29, 71, 72, 82, 89, 93, 101, 118
individuals, 77, 106, 107, 108, 114
induction, 4, 5, 9, 10, 12, 13, 16, 17, 19, 21, 23, 29, 30, 34, 46, 49, 50, 51, 52, 54, 55, 56, 110
information processing, 71, 72, 74, 76, 90, 94
inhibition, 12, 15, 23, 26, 27, 48, 50, 67, 68, 70, 72, 76, 83, 90, 108, 122
injury/injuries, vii, ix, 100, 110, 111, 121

interface, 6, 18, 30, 31, 113
interneuron(s), 47, 62, 63, 68, 69, 70, 71, 76, 83, 108, 123

K

Keynes, 4, 47, 50

L

laminar, viii, 68, 87, 91, 99, 100, 102, 105, 123
learning, 62, 72, 73, 74, 75, 78, 83, 84, 85, 108, 109, 111, 114, 116, 122, 123
lesions, 70, 87, 88, 92, 110
ligand, 15, 23, 56
lissencephaly, vii, ix, 100, 103, 107, 113, 115, 120, 121
locus, 18, 19, 31, 32, 34, 36, 86, 100, 102, 108

M

mAChRs, 67, 68, 72, 73, 74, 75, 76, 77
mammals, 23, 24, 39, 84, 101, 111
manipulation, 29, 106, 109
medial prefrontal cortex, 65, 79, 85
memory, 62, 72, 75, 77, 78, 83, 85, 88, 95, 108, 109, 112
Meynert basal magnocellular nucleus, 63
mice, 18, 19, 20, 21, 23, 31, 32, 34, 45, 71, 95, 96, 100, 102, 103, 105, 106, 107, 109, 110, 111, 112, 114, 116, 118, 119, 121, 122, 123
midbrain, 2, 4, 6, 17, 31, 32, 53, 56
migration, 10, 11, 100, 102, 103, 104, 110, 113, 115, 117, 122
models, 35, 37, 52, 85, 116
molecules, 11, 21, 27, 102
morphology, 7, 29, 32, 36, 102, 103

motif, 19, 20, 33, 101
motor control, viii, 62, 65, 104, 106, 108
motor function, 100, 114
motor impairments, 100, 105
mPFC, 65, 66, 67, 74, 75, 76
muscarinic receptor, 67, 71, 75, 79, 82, 85, 90
mutant, viii, 31, 33, 36, 99, 100, 101, 102, 105, 109, 113, 116, 120, 121, 123
mutation(s), 35, 36, 45, 51, 112, 100, 103, 104, 107, 115, 118, 123

N

nAChR(s), 67, 68, 70, 72, 73, 75, 76, 77
National Academy of Sciences, 82, 83, 84, 88
neocortex, 64, 65, 66, 68, 70, 71, 77, 83, 86, 121
nervous system, 4, 5, 6, 48, 50, 55
neural, vii, 1, 2, 4, 5, 6, 8, 9, 11, 12, 13, 16, 17, 23, 24, 27, 29, 30, 34, 35, 40, 41, 42, 43, 44, 45, 46, 48, 50, 51, 52, 54, 55, 56, 60, 70, 71, 76, 79, 81, 84, 86, 88, 89, 95, 96, 103, 104, 106, 117, 118
neural development, 6, 27, 35, 45
neurogenesis, 37, 49, 123
neurological disease, vii, ix, 100
neurons, 49, 54, 62, 63, 65, 66, 67, 68, 69, 70, 71, 73, 75, 76, 77, 80, 81, 82, 83, 84, 85, 87, 89, 90, 91, 92, 93, 95, 100, 102, 103, 105, 106, 112, 121
nicotine, 68, 82, 89
nicotinic receptor, 67, 68, 79, 81
notochord, 6, 9, 52
nucleus/nuclei, viii, 15, 41, 61, 63, 64, 65, 66, 74, 75, 77, 80, 81, 88, 90, 96, 100
nucleus basalis magnocellularis, 63, 90
null, 31, 102, 104, 107

P

Parkinson, 62, 77, 80, 81, 86, 90
parvalbumin, 68, 82, 108
pathway(s), vii, viii, 5, 6, 12, 15, 22, 23, 26, 27, 28, 35, 36, 43, 44, 62, 64, 65, 82, 87, 88, 91, 96 102, 118
patterning, vii, 1, 2, 3, 4, 5, 6, 9, 10, 11, 12, 13, 17, 20, 23, 24, 30, 33, 34, 35, 37, 38, 39, 40, 41, 42, 43, 44, 45, 46, 47, 48, 50, 51, 53, 54, 55, 57
phenotype(s), viii, 31, 36, 50, 99, 100, 101, 102, 104, 109, 120, 121
phosphorylation, 15, 102, 111, 115
plasticity, ix, 59, 62, 68, 71, 72, 74, 90, 92, 95, 100, 108, 110, 111, 112, 115, 116, 119, 120, 121, 122, 123
population, 10, 47, 62, 68, 69, 70, 77, 110
prefrontal cortex, 65, 75, 79, 81, 84, 85, 88, 108, 111
primary visual cortex, 73, 74, 91
principles, vii, 1, 3
processed, 21, 75, 116
progenitor cell, 5, 23, 110, 113
proliferation, 2, 5, 12, 28, 32
proteasome, 12, 15, 23
proteins, 5, 15, 17, 18, 21, 23, 24, 26, 28, 37, 47, 49, 53, 67, 86, 112, 113, 114
psychiatric diseases, 100, 111
psychiatric disorders, vii, ix, 100, 107, 108

R

receptor(s), viii, 15, 22, 42, 51, 58, 67, 68, 71, 74, 75, 76, 77, 78, 79, 81, 82, 85, 86, 87, 90, 91, 96, 99, 101, 102, 104, 111, 112, 113, 118, 122, 123
recovery, 110, 115, 118, 119
reeler, viii, 99, 100, 101, 102, 103, 104, 105, 106, 107, 108, 110, 111, 112, 113,

Index

114, 116, 117, 118, 119, 120, 121, 122, 123
reelin, v, vii, viii, 99, 100, 101, 102, 103, 104, 105, 107, 108, 109, 110, 111, 112, 113, 114, 115, 116, 117, 118, 119, 120, 121, 122, 123
regionalization, 3, 5, 6, 9, 12, 17, 41, 50, 55, 57
REM, 62, 69, 70, 72, 86
repression, 13, 15, 16, 40, 42
repressor, 13, 15, 23, 49
response, 17, 21, 23, 28, 33, 43, 55, 60, 68, 73, 74, 75, 79, 87, 89, 90, 116
rhythm, 70, 89, 92
rodents, viii, 47, 62, 64, 65, 66, 67, 73, 74, 81, 100

S

schizophrenia, 62, 90, 107, 108, 109, 112, 113, 114, 115, 116, 117, 120, 121, 122
schizophrenic patients, 119, 120, 122
secretion, 6, 10, 21, 22, 23, 49, 56, 103, 118
segregation, 2, 10, 25, 26, 28, 29, 30, 36, 37, 41
sensory cortices, 65, 66, 67, 75, 81, 86, 96
septum, 63, 64, 70, 85
serine, 15, 101, 119
shape, 27, 29, 36, 76
signal transduction, viii, 2, 3, 22, 23, 36
signaling pathway, vii, viii, 1, 3, 5, 6, 9, 23, 28, 35, 44, 101, 102, 103, 105
signalling, 40, 43, 45, 47, 51, 52, 53, 54, 56, 82
signals, 5, 6, 10, 15, 16, 17, 33, 37, 42, 45, 48, 50, 52, 58, 83, 115
slow wave sleep, 69
sonic hedgehog, 2, 40, 43, 44, 46, 47, 49, 51, 52, 56, 57
species, viii, 1, 4, 12, 15, 19, 24, 25, 29, 30, 31

spinal cord, 4, 39, 48
spine, ix, 100, 117, 120
state(s), 80, 87, 88, 92
stimulation, 67, 70, 72, 73, 74, 76, 81, 83, 87, 90, 95, 96, 106
stimulus, 74, 76, 96
striatum, 75, 88, 112
stroke, 115, 116, 117, 123
structure, viii, 2, 10, 36, 57, 63, 70, 79, 99, 103, 105
substantia innominata, 63, 64, 66, 78, 81
substrate(s), 81, 100, 110
suppression, 85, 88, 92
survival, vii, 1, 2, 3, 12, 15, 16, 50, 60, 105
SWS, 69, 70
symptoms, 107, 111, 117
synaptic plasticity, 71, 72, 90, 92, 108, 110, 112, 120, 122, 123
synchronization, 71, 83, 90, 92
syndrome, 36, 39, 43, 49, 105, 107, 117

T

target, 15, 18, 23, 41, 62, 106, 111
telencephalon, 2, 8, 11, 46, 63
territory, 3, 11, 12, 13, 16, 17, 18, 26, 30, 31, 32, 33, 36, 37, 57
testing, 107
thalamus, vii, 1, 2, 3, 7, 9, 10, 11, 13, 16, 18, 25, 26, 29, 30, 31, 33, 34, 35, 36, 37, 46, 47, 54, 55, 63, 64, 75, 87, 88
that, vii, viii, 1, 2, 3, 4, 5, 6, 7, 9, 10, 11, 12, 13, 16, 18, 19, 20, 21, 22, 23, 25, 26, 27, 28, 29, 30, 31, 32, 33, 35, 36, 37, 42, 44, 47, 52, 61, 62, 64, 65, 66, 67, 68, 69, 70, 71, 72, 73, 74, 75, 76, 77, 78, 93, 99, 101, 102, 103, 105, 106, 107, 108, 109, 110, 111
theta oscillations, 70, 72, 76, 80, 84, 93
tissue, 5, 6, 10, 17, 30, 33, 43
tonic, 69, 73, 74, 82, 89

training, 109, 110, 118
transcription, 3, 15, 16, 20, 40, 46, 54
transcription factors, 3, 15, 46, 54
transduction, viii, 2, 3, 22, 102
transmission, 65, 78, 82, 89, 90, 91, 92
transport, 15, 22, 36, 45, 47
treatment, 85, 111, 116
turnover, 41, 108, 112

U

United States (USA), 58, 82, 83, 84, 88

V

vertebrates, 2, 4, 20, 22, 23, 24, 40, 41

W

Wnt signaling, 5, 8, 12, 21, 52, 56

Z

zinc, 22, 30, 47